The Anthropocene: Politik—Economics—Society—Science

Volume 28

Series editor

Hans Günter Brauch, Mosbach, Germany

More information about this series at http://www.springer.com/series/15232
http://www.afes-press-books.de/html/APESS.htm
http://www.afes-press-books.de/html/APESS_28.htm

Saleemul Huq · Jeffrey Chow
Adrian Fenton · Clare Stott
Julia Taub · Helena Wright
Editors

Confronting Climate Change in Bangladesh

Policy Strategies for Adaptation and Resilience

Editors
Saleemul Huq
Climate Change Group
International Institute for Environment and
Development
London, UK

International Centre for Climate Change and
Development
Dhaka, Bangladesh

Jeffrey Chow
International Centre for Climate Change and
Development
Dhaka, Bangladesh

Adrian Fenton
International Centre for Climate Change and
Development
Dhaka, Bangladesh

Clare Stott
International Centre for Climate Change and
Development
Dhaka, Bangladesh

Julia Taub
International Centre for Climate Change and
Development
Dhaka, Bangladesh

Helena Wright
International Centre for Climate Change and
Development
Dhaka, Bangladesh

For more on this book, see: http://www.afes-press-books.de/html/APESS_28.htm

ISSN 2367-4024   ISSN 2367-4032 (electronic)
The Anthropocene: Politik—Economics—Society—Science
ISBN 978-3-030-05236-2   ISBN 978-3-030-05237-9   (eBook)
https://doi.org/10.1007/978-3-030-05237-9

Library of Congress Control Number: 2018962775

Copyediting: PD Dr. Hans Günter Brauch, AFES-PRESS e.V., Mosbach, Germany

This Springer imprint is published by the registered company Springer Nature Switzerland AG.
The registered company address is: Gewerbestrasse 11, 6330 Cham, Switzerland

Acknowledgements

The authors would like to express their gratitude to all chapter authors for taking the time to submit their work to the contribution to this collective volume. We would also like to thank the staff of the *International Centre for Climate Change and Development* (ICCCAD) and the Gobeshona initiative in Dhaka, Bangladesh, for support in the coordination of the chapter submission process. Finally, we would like to express our thanks to all the anonymous peer reviewers who took their time to peer review the chapters within their book. The book could not have been possible without the time and expertise of all those who have helped through the whole publication process.

The cover photograph was taken by Sarah Goldstein Opasiak who also granted the permission to use it here.

Contents

List of Figures

List of Tables

Chapter 1
Introduction: Bangladesh Responds to Climate Change

Helena Wright, Adrian Fenton, Saleemul Huq, Clare Stott, Julia Taub and Jeffrey Chow

Abstract Bangladesh is a country that is highly vulnerable to the impacts of climate change, and a broad range of practices have emerged to adapt to these impacts. The book presents a range of sectors that are affected by the impacts of climate change, such as agriculture, water and health, as well as covering thematic areas relating to responses to climate change, such as governance and finance, communication and gender. Measures to adapt to climate impacts in the agricultural sector range from hard engineering measures like construction of polders, to soft socio-economic measures, such as changes in cropping patterns. In the water sector, non-structural approaches to risk reduction include community-based disaster management initiatives. Across all practice areas there are barriers and challenges to confronting the impacts of climate change, including knowledge gaps. The chapters of this book emerged as part of the Gobeshona initiative in Bangladesh, a knowledge sharing platform for climate change research on Bangladesh.

Keywords Bangladesh · Adaptation · Climate · Impacts · Poverty

Bangladesh is a low-lying agrarian country located in the Ganges-Brahmaputra Delta, which is highly vulnerable to the impacts of climate change. A broad range of practices have been identified to adapt to these impacts, some of which are already taking place. The chapters of this book, which emerged out of the

Dr. Helena Wright, International Centre for Climate Change and Development (ICCCAD), Dhaka, Bangladesh, Corresponding Author, e-mail: drhelenawright@gmail.com.

Dr. Adrian Fenton, International Centre for Climate Change and Development (ICCCAD), Dhaka, Bangladesh.

Dr. Saleemul Huq, Director, International Centre for Climate Change and Development (ICCCAD), Dhaka, Bangladesh.

Clare Stott, International Centre for Climate Change and Development (ICCCAD), Dhaka, Bangladesh.

Julia Taub, Project Officer, Global Network of Civil Society Organisations for Disaster Reduction (GNDR), London.

Dr. Jeffrey Chow, International Centre for Climate Change and Development (ICCCAD), Dhaka, Bangladesh.

© Springer Nature Switzerland AG 2019   1
S. Huq et al. (eds.), *Confronting Climate Change in Bangladesh*,
The Anthropocene: Politik—Economics—Society—Science 28,
https://doi.org/10.1007/978-3-030-05237-9_1

Gobeshona initiative, present a range of sectors affected by climate change as well as thematic areas relating to responses to climate change in Bangladesh. In particular, agriculture is a sector which is affected by increased temperature and unpredictable rainfall patterns in Bangladesh (Mondal et al. 2012). As explained in Chap. 2, the agricultural sector employs around half of the civilian workers in Bangladesh and plays an important role in poverty reduction. The range of measures to adapt to climate impacts in the agricultural sector identified by Mondal et al. in this volume range from hard engineering measures like construction of polders, to soft socio-economic measures, such as changes in cropping patterns. Mondal et al. (2012) find that the majority of the identified adaptation practices in agriculture are for the purpose of implementing actions which reduce hazards, enhance production, and bring direct, tangible benefits. However, the agriculture chapter outlines a number of barriers in relation to these practices including a lack of capital, lack of access to resources, lack of knowledge and information, and lack of institutional capacity. In the coastal zone, the government is the primary provider of the identified agricultural adaptations, while Mondal et al. (2012) explain that the role of the private sector as an adaptation provider is not studied or well documented within adaptation literature.

The water sector and availability of freshwater both face the threat of loss and damage due to current and expected climate change impacts. As outlined in Chap. 3, Bangladesh's population of nearly 160 million relies heavily upon its hydrological systems and is therefore vulnerable to events that will be exacerbated by climate change, such as floods, storm surges, drought, sea level rise, and salinity intrusion, as well as loss of fish spawning grounds and reductions in agricultural production due to changes in the hydrological regime. As Mukherjee et al. explain in Chap. 3, the wetland systems of Bangladesh – known as "haor" regional rivers are susceptible to flash flooding which can affect harvests, and during the 2004 flood, more than two thirds of the boro production was lost due to an early flash flood event coinciding with the harvest (CEGIS 2012). However, while efforts have been made to adapt, it is noted in this volume that some initiatives may be maladaptive despite having claimed benefits over the short term. For instance, while river embankments have been constructed as adaptation measures these can also restrict the sediment inflow to the flood plain, reducing the nutrient availability of the topsoil (Brouwer et al. 2007). Mukherjee at al. explain that other non-structural approaches to risk reduction include early warning systems and community-based disaster management initiatives.

In Bangladesh, forests and wetlands are at risk from climate impacts, but protecting forests can also be considered as an opportunity to respond to climate change. As outlined in Chap. 4, mangrove forests are coastal forests that can help reduce damage from sea level rise and erosion (FAO 2007). However, as detailed by Chow et al. (in this volume) greenbelts can only provide the role of protecting against intense storms if they are appropriately designed and managed. The forest chapter describes major adaptation initiatives which have taken place for the conservation of the Sundarbans forest, which is the largest contiguous mangrove ecosystem in the world and is a UNESCO World Heritage Site. As explained by

Chow et al. (in this volume), many important knowledge gaps remain that require continued investigation; for example, we do not know the width of mangrove plantation that will provide adequate storm protection, as currently these plantation widths are set almost arbitrarily. This demonstrates that further research is required to enhance the impacts of forests and wetlands for adaptation. Relatedly, plantation of mangroves can be considered as an example of *ecosystem-based adaptation* (EbA), which refers to the use of biodiversity and ecosystem services as part of a climate change adaptation strategy, as examined in Chap. 5. Among other ecosystem services, Saroar et al. explain in this volume that mangroves can provide a line of defence against cyclonic storm surges, erosion, and salinity intrusion for coastal communities, infrastructure, and livelihood assets, and can also provide pollution control, water purification, and improvement of drainage.

The chapter on Governance and Finance in this volume explains that in the initial stages, policy concerns for environmental protection in Bangladesh were reflected for the first time in the Fourth Five Year Plan (1990–1995), and have been included in other five-year plans since then. Since the signing of the Kyoto Protocol, Pervin et al. (in this volume) explain that the governance of climate change in Bangladesh has been characterised by specific interventions, such as the creation of policy provisions, research and technological innovations, and the establishment of funding entities. Bangladesh was the one of the few *Least Developed Countries* (LDCs) to develop a National Adaptation Programme of Action (NAPA) in 2005, and this was then revised in 2009. The Government of Bangladesh also unveiled the *Bangladesh Climate Change Strategy and Action Plan* (BCCSAP) in 2008. The chapter describes the various funding windows in Bangladesh including the *Bangladesh Climate Change Trust Fund* (BCCTF) established in 2009, the *Bangladesh Climate Change Resilience Fund* (BCCRF) established in 2010, and the *Pilot Programme for Climate Resilience* (PPCR).

Chapter 7 explains that media and education systems can play an important role in alerting and preparing people for climate-related disasters as well as empowering them to minimise risks, which means that raising public awareness and under-standing of climatic risks through disseminating accurate information is an important part of building long-term resilience. At the global level, the Sendai Framework for Disaster Risk Reduction emphasises the utilisation and strength-ening of all kinds of media to support successful disaster risk communication. The chapter on media and communication provides examples of different forms of communication interventions which are relevant to adaptation in Bangladesh. For example, in 2007 before the powerful Cyclone Sidr struck Bangladesh, the Bangladesh Government began to broadcast warnings five days in advance on radio and television (Paul/Dutt 2010). Emergency evacuation orders were also issued almost twenty-seven hours before landfall of the cyclone, which the authors argue helped reduce the death toll. The Government of Bangladesh has recognised the role of the media in the National Broadcasting Policy of 2014 (GoB 2014), which makes a provision for telecasting emergency weather bulletins and producing cli-mate change awareness programmes, as Afroz et al. (in this volume) explain in this volume. Chapter 7 explains that there is a wealth of traditional knowledge that has been preserved for generations alongside locally adaptive mechanisms for survival, so this local knowledge can be being integrated within media interventions to appropriately support resilience.

Gender inequalities in Bangladesh often mean that women are more affected by climate change than male counterparts. The gender chapter highlights a few of these gender inequalities from the perspective of rights. It begins with a boxed text outlining international gender and climate change policy, followed by a section on the Bangladesh context and specific gender inequalities and discriminations that contribute to women's vulnerability. For example, a study by Neumayer/Plümper (2007) found that women and children are 14 times more likely to die or be injured in a disaster than men. Reggers explains in Chap. 8 that the decision-making over resource use, as well as buying and selling of land before or after climate-related events often rests with men. Finally, Reggers notes that women's relative lack of mobility in public spaces can result in women not receiving early warning signals before cyclones. The author finishes by providing examples of national and community level efforts to address the gender dimensions of climate change.

The health sector is another sector particularly affected by climate change in Bangladesh. The Intergovernmental Panel on Climate Change (IPCC 2014), cited in Chap. 9, affirms that "recent decades have seen warming air and ocean temperatures, changing rainfall patterns, variations in the frequency and intensity of several extreme events including droughts, floods and storms and rising sea levels" and that the changing climate will adversely affect the health of human populations. The chapter explains the three basic pathways by which climate change affects health: primary or direct impacts (such as heat stress), secondary or indirect impacts (such as shifts in diseases), and tertiary or long-term implications (mediated by social, political and economic systems). Rahaman et al. argue in this volume (Chap. 9) that the increasing frequency of extreme climatic events such as floods can affect health infrastructure like hospitals, clinics and healthcare facilities, while climatic events that result in reduced food production can affect nutrition due to the declining food quality and quantity. As for other impacts, women, children and the elderly are often particularly vulnerable.

Chapter 10 explains that climate-related events such as cyclones, land erosion and flooding can also contribute to displacement and human migration. In the context of Bangladesh this can either lead to increased risk or create new opportunities. As Saha et al. explain in this volume, those who are forced to leave their homes often take shelter in places where they lack legal rights or social services, but there are also examples where settling in a new location offers opportunities for coping and resilience. Internal displacement and migration in Bangladesh is also linked to the movement of people to cities. Urban areas are also expanding in Bangladesh, partly in response to rural environmental and climate hazards. Taking the case of Dhaka, Chap. 11 finds that flooding poses a significant threat and the city lacks proper mechanisms to counter flooding, due to the combination of Dhaka's climatic and geographical conditions, as well as weak political and economic capacity to address the situation. Unregulated housing developments in low-lying areas can exacerbate the problem by preventing natural drainage

(Morshed 2013). The authors identify solutions such as the need for flood warning systems and well-managed and equipped flood shelters.

Finally, in the energy sector, solar home systems have emerged as a widespread technology enabling electrification to reach remote rural communities. As Muzammil and Ahmed explain in Chap. 12, the Solar Home System (SHS) programme in Bangladesh has grown to be one of the largest off-grid electrification initiatives in the world, and has also been described by the World Bank as the fastest growing SHS programme in the world (World Bank 2014). The final chapter reflects on the benefits of solar home systems for low carbon, resilient development opportunities in Bangladesh, as well identifying emerging challenges and lessons for other countries. It is noted that the SHS programme in Bangladesh benefitted largely from a strong pre-existing network of competitive microfinance institutions with a broad reach in rural areas (Sadeque et al. 2014). Overall, Muzammil and Ahmed note that increased support from donors allowed the programme to reach the poor, as well as enhance market development and catalyse finance for smaller players, providing an innovative funding model with lessons for other countries.

The chapters of this book all emerged as part of the Gobeshona initiative in Bangladesh, a knowledge sharing platform for climate change research on Bangladesh that aims to bring together the national and international research community to encourage sharing, enhance research quality and make climate change research on Bangladesh more effective (ICCCAD 2017). Some chapters address sectors that are particularly vulnerable to climate change, including agricultural, coastal zones, water, while others address particular thematic issues of relevance to adaptation in Bangladesh, including governance, communication and gender. The aim is to connect researchers with other stakeholders, with the hope that the resulting publications can be used in response to climatic impacts in Bangladesh.

References

Brouwer R, Akter S, Brander L, Haque E (2007). Socioeconomic vulnerability and adaptation to environmental risk: a case study of climate change and flooding in Bangladesh. Risk Analysis 27: 313–326.

Center for Environmental Geographic Information Systems (CEGIS) (2012). Master plan of haor areas. Dhaka: Bangladesh Haor and Wetland Development Board, Ministry of Water Resources, Government of the Peoples Republic of Bangladesh.

Food and Agriculture Organization of the United Nations (FAO) (2007). The World's Mangroves 1980–2005. FAO Forestry Paper 153. FAO, Rome, Italy.

GoB (2014). Bangladesh Gazette. Bangladesh Government Press, Dhaka, Bangladesh.

IPCC (2014). Climate Change 2014: Impacts, Adaptation, and Vulnerability. Contribution of Working Group II to the Fifth Assessment Report of the Intergovernmental Panel on Climate Change. Cambridge University Press, Cambridge, United Kingdom and New York, NY, USA, pp. 1039–1099.

International Centre for Climate Change and Development (2017). Available at: http://www. icccad.net/gobeshona/, accessed on 29 November 2017.

Mondal MS, Islam AKMS, Modhu MK (2012). Spatial and temporal distribution of temperature, rainfall, sunshine and humidity in context of crop agriculture. Comprehensive Disaster Management Program, Ministry of Food and Disaster Management, Dhaka.

Morshed MM (2013). Detailed Area Plan (DAP): Why It Does Not Work? Planned Decentralization: Aspired Development: Souvenir published on World Habitat Day. Available at: http://www.bip.org.bd/SharingFiles/journal_book/20140128161651.pdf.

Neumayer E, Plümper T (2007). The gendered nature of natural disasters: the impact of catastrophic events on the gender gap in life expectancy, 1981–2002. Annals of the Association of American Geographers, 97(3), 551–566.

Paul BK, Dutt S (2010). Hazard warnings and responses to evacuation orders: the case of Bangladesh's cyclone Sidr. Geographical Review 100 (3):336–355.

Sadeque Z, Rysankova D, Elahi R, Soni R. (2014). Scaling up access to electricity: the case of Bangladesh. Live Wire 2014/21, World Bank Group. Available at: https://openknowledge. worldbank.org/bitstream/handle/10986/18679/887020BRI0Live00Box385194B00PUBLIC0. pdf?sequence=1, accessed on 6 September 2016.

World Bank (2014). http://www.worldbank.org/en/news/press-release/2014/06/30/bangladesh-receives-usd-78-million-to-install-an-additional-480000-solar-home-systems, accessed on 6 September 2016.

Chapter 2
Agricultural Adaptation Practices to Climate Change Impacts in Coastal Bangladesh

**M. Shahjahan Mondal, Mohammad Towheedul Islam,
Debanjali Saha, Muhammad Shahriar Shafayet Hossain,
Prodip Kumar Das and Rezaur Rahman**

Abstract Bangladesh is an agrarian country and about one-third of its cultivable lands are in coastal and offshore areas which are highly vulnerable to climate change impacts. To adapt to such impacts, a number of policies, plans and adaptation measures have been suggested. However, it is not clear how many of these potential adaptation measures are actually in practice. In this study, we carry out an inventory of agricultural adaptation practices in the coastal zone of Bangladesh and present a synthesis of the inventory. The inventory is developed by recording multiple dimensions of adaptations. It records the purpose, geographic location, provider/beneficiary, timing, drivers, barriers to adaptation, gender aspects and sustainability issues of the adaptation practices. The findings of the study indicate that about 85 agricultural adaptations are now in practice, the majo`rity of which are infrastructural-technological in scope. Almost all the adaptations are deliberate actions which come with a tangible aim of taking action/implementing change. The majority of the adaptations have been in response to long-term chronic stresses, such as salinity. Lack of capital, access to resources, knowledge and information, and centralised decision-making process appear to be some of the barriers to taking up the adaptations. The current evidence suggests that Bangladesh has embraced a

M. Shahjahan Mondal, Institute of Water and Flood Management, Bangladesh University of Engineering and Technology, Dhaka-1000, Bangladesh, Corresponding Author, e-mail: mshahjahanmondal@iwfm.buet.ac.bd.

Mohammad Towheedul Islam, Department of International Relations, University of Dhaka, Bangladesh and Refugee and Migratory Movement Research Unit, University of Dhaka, Bangladesh.

Debanjali Saha, Institute of Water and Flood Management, Bangladesh University of Engineering and Technology, Dhaka, Bangladesh.

Muhammad Shahriar Shafayet Hossain, Institute of Water and Flood Management, Bangladesh University of Engineering and Technology, Dhaka, Bangladesh.

Prodip Kumar Das, Refugee and Migratory Movement Research Unit, University of Dhaka, Bangladesh.

Rezaur Rahman, Institute of Water and Flood Management, Bangladesh University of Engineering and Technology, Dhaka, Bangladesh.

© Springer Nature Switzerland AG 2019   7
S. Huq et al. (eds.), *Confronting Climate Change in Bangladesh*,
The Anthropocene: Politik—Economics—Society—Science 28,
https://doi.org/10.1007/978-3-030-05237-9_2

mix of adaptations for agricultural development, though there are still areas for improvement.

Keywords Adaptation practice · Agriculture · Coastal area · Adaptation inventory Gaps between policy and practice

2.1 Introduction

Bangladesh is a small country with a large population of about 160 million. About 32% of its population live below the upper poverty line and 18% below the lower poverty line (BBS 2013). The per capita net cropped area available in the country is less than 0.05 ha. Bangladesh is an agrarian country, and about 53% of households have farmland holdings and 48% of civilian workers are employed in the agriculture sector (BBS 2011, 2013). Being a predominantly deltaic country, about 57% of its total area is cultivable. There are three overlapping crop seasons (Rabi, Kharif-I and Kharif-II) and rice is the single most dominant crop cultivated in all these seasons, with about 77% coverage of the total cropped area. Jute, potato, wheat, maize, oilseeds, pulses, vegetables and spices are among the major non-rice crops. Agriculture plays an important role in rural employment and poverty reduction. Irrigated agriculture in particular plays a significant role in enhancing agricultural production and income, which contribute to the reduction of poverty and improvement in living conditions (Saleh/Mondal 2009). However, the agriculture sector is highly vulnerable to climate change impacts induced by global warming. Increased temperature, erratic and unpredictable rainfall, shortening of winter, foggy weather, sea level rise, increased flooding, and increased cyclones and storm surges are among the hydro-climatic factors which affect the country's agriculture sector (Mondal et al. 2012).

About one-third of the cultivable land in Bangladesh lies in the coastal and offshore areas. These areas are even more vulnerable to the effects of climate change due to their proximity to the sea, low land topography, the confluence of the mighty Ganges-Brahmaputra-Meghna rivers, reduced freshwater supplies from upstream, and fragile agro-ecosystems. About 37% of the cultivable coastal land is already affected by varying degrees of soil salinity (Mondal et al. 2015). The spatial extent of such area is increasing over time. Between 1973 and 2009, the salt-affected coastal and offshore areas increased by about 27% (SRDI 2012). Apart from salinity, cyclones, storm surges, tidal flooding, waterlogging and erosion are among the major hazards in the coastal areas. The frequency and intensity of these hazards are expected to increase, and their adverse impacts are likely to intensify under the changing climatic and environmental conditions in the future. In addition, the relatively high incidence of poverty and the high proportion of people dependent on ecosystems for their livelihoods further add to the vulnerability of coastal people to climate change impacts. To mitigate the adverse effects of climate change, adaptation is seen as a practical strategy in both policy and practice.

A number of studies have been conducted on agricultural adaptations in Bangladesh. However, most of these studies have been on potential adaptation options rather than on adaptation practices. For example, different kinds of studies were recommended in Brammer et al. (1993) to improve knowledge about the probable impacts of climate change on plant growth and yield, and to monitor environmental changes as they occur. Adoption of hybrid rice, development of pest and disease-resistant, stress-tolerant and short-duration crop cultivars, and development of crop varieties with low-transpiration ratios were postulated as potential adaptation options under a changed climate (Karim et al. 1998). Awareness building, infrastructural development, disaster preparedness, post-disaster rehabilitation, crop varietal development, improvement in agricultural management practices, development of agricultural extension services, promotion of agro-processing techniques, market infrastructure development, improved irrigation and water management, and integrated coastal zone management were identified as adaptation options by Karim (2011). Seventeen rice varieties have been developed in Bangladesh to withstand adverse climatic and environmental conditions (Rabbani et al. 2015). Two of them are submergence-tolerant, nine are salt-tolerant and the remaining varieties are drought-resistant. Other studies on potential adaptation options relate to crop agriculture (Ericksen et al. 1993; Ali 1999; World Bank 2000, 2011; Faruque/Ali 2005; Ahmed 2006; Asaduzzaman et al. 2010; Rawlani/Sovacool 2011; Asia Foundation 2012; Dev 2013). In contrast, there are relatively fewer studies on adaptation practices, i.e., on evidence of adaptations (Younus et al. 2005; Oxfam International 2009; Ahmed 2010; Sterrett 2011; Abedin/Shaw 2013). Also, there is no systematic study on such practices in Bangladesh from a holistic perspective, i.e., answering questions like which groups are adapting and benefiting, what have been the major drivers of adaptation, what are the ultimate aims of adaptations, in which geographical locations are adaptations occurring, whether the gender dimension is taken into account, etc. These holistic studies are also not available in the international arena with the exception of the one done for the UK by Tompkins et al. (2010). Such a study would help inform where Bangladesh stands in terms of adapting to climate change, identifying successful and unsuccessful adaptations, and designing better climate change policies and actions in the future.

2.2   Methodology

This study was conducted in the coastal zone of Bangladesh, as it represents the most vulnerable part of the country. Since there are a number of definitions of climate change adaptation, we draw on the IPCC (2014) definition: "The process of adjustment to actual or expected climate and its effects. In human systems, adaptation seeks to moderate harm or exploit beneficial opportunities". The study looks at adaptation to climate stressors, which include exposure to sudden-onset shocks, such as floods, and/ or to slow-onset stresses, such as changes in temperature and rainfall. Adaptation in this study also refers only to adjustments made by humans.

Following this definition, an inventory of adaptation practices in the coastal zone of Bangladesh was prepared in order to collate the literature. To prepare the inventory, a protocol was first developed to collect evidence of adaptation to climate change. Currently observed and documented agricultural adaptations were within the purview of the protocol. The experience in developing the UK inventory of observed adaptations (Tompkins et al. 2009) helped in the formulation of the protocol for this inventory. Though agriculture broadly encompasses crop, livestock, forestry and fishery sectors in some countries, only crop agriculture was considered in this study, as the agriculture ministry in Bangladesh is responsible solely for the crop sector and there are separate ministries and statistics available for others. Published literature – both peer-reviewed and grey – was collated in a universal spreadsheet template. No distinction was made between peer-reviewed and grey literature to qualify for the inventory except that the grey literature must contain authors/institutions, content with evidence of adaptation and year of publication. The template had 43 columns recording inter alia the purpose, geographic location, provider/beneficiary, timing, drivers, barriers to participation, gender issues, current and future damaging aspects, and long-term sustainability of the adaptation practices.

We collected literature available in both printed and electronic copies. Printed literature available from different national government agencies, non-governmental organisations, and academic and research institutes was collected. *The Bangladesh Climate Change Trust Fund* (BCCTF) has funded 219 projects since 2009–10 and the *Bangladesh Climate Change Resilience Fund* (BCCRF) has funded 13 projects. The documents available on these projects were collected and included in the inventory. For electronic literature, we searched in a number of online sources including academic databases (ScienceDirect, Springer, Wiley, OARE, etc.) and organizational web pages. The databases were searched with a number of relevant keywords, such as *adapt, resilien, vulnerab, climat, chang, variab, extreme, weather, disaster, risk, cop, recover, rehabilitat, repair, migrat, displac, hazard, insurance,* etc. and their appropriate combinations using Boolean and truncation operators. In addition, formal letters were sent to about 30 government, non-government and private organisations with a request to share available adaptation related documents. After sending the letters, the organisations were also contacted over telephone or in person. A few more documents were received through these processes and included in the inventory. One limitation of this approach was that only those practices which have been documented and defined as 'adaptations' were qualified to be included in the inventory. Many autonomous or spontaneous adaptations could be occurring in the coastal zones without proper documentation. Also, we did not attempt to identify to what extent the adaptations were in reaction to climatic impacts caused by global climate change or by other contextual factors, as that was beyond the scope of the study.

A half-day national workshop was arranged in Dhaka in 2015 to share and validate the findings of the inventory. The workshop was well attended by adaptation practitioners, professionals, implementers, academics, and researchers from different government and non-government organisations.

2.3   Analysis of Adaptation Practices

The inventory that we have prepared records many aspects of adaptation practices. These aspects were analysed using simple statistics, such as frequency of the practices, and are presented in this section. After describing the different adaptations that are in practice and their spatial distribution in the next two subsections, we will discuss the different forms, drivers, providers and beneficiaries of the adaptations, and barriers to the adaptation practices. Gender issues as reflected in the adaptation documents, and the various adaptation-linked policies and plans as formulated in the country, are highlighted in the final two subsections. The statistics are presented in terms of number of practices rather than by number of documents.

2.3.1   Adaptation Practices

About 85 different adaptations related to crop agriculture were identified as being currently practiced in the coastal region of Bangladesh. These range from hard engineering measures such as construction of polder, to soft socio-economic measures such as changes in cropping pattern, to technological innovations such as cultivation of stress-tolerant crops. Of the total practices, 53 are found to be infrastructural-technological in scope and 31 are socio-economic. Infrastructural adaptations typically characterise physical structures such as polders, regulators, sluices, culvert, canals, storage reservoirs, tube wells, pumps, and revetments; while technological adaptations include farm machineries, crop varietal development, irrigation canal lining, and new irrigation techniques.

The major adaptation practices have been the construction of polders to protect agricultural lands from flooding, construction of drainage infrastructures to alleviate waterlogging, innovation in crop technologies such as stress tolerant and short duration crop varieties, agricultural mechanisation to ease post-harvest activities and reduce operation time between successive crops, and introduction of integrated farming practices and cropping pattern changes. In addition to these practices, innovations in vegetable gardening such as floating beds and hanging gardens, rainwater harvesting for irrigation, dyke cropping to incorporate vegetables and other crops in a single season, and relay cropping were among the emerging adaptation practices. Some research in innovative agricultural practices, training and knowledge dissemination, and policy formulation were also found among the agricultural adaptation practices.

2.3.2 Spatial Distribution of the Practices

There are 19 districts in the coastal area, all of which have witnessed at least some agricultural adaptations. However, there are regional variations in adaptation practices (Fig. 2.1). For example, more adaptations were identified in the exposed 13 districts than the interior 8 districts. Also, the western Khulna-Satkhira region has witnessed more adaptations than the central Barisal-Patuakhali region due to strong NGO (non-governmental organisation) activities after cyclones Sidr (in 2007) and Aila (in 2009). Since the salinity problem is greater in the former region, salinity-related adaptations are also more numerous there. The moderate number of adaptations in the eastern Chittagong district are mainly infrastructural in nature, such as construction of pumping stations, irrigation canals, polders, sluice gates, regulators and culverts.

2.3.3 Forms of Adaptation Practices

Agricultural adaptations in the coasts and offshore islands have taken place in various forms. About 27% of the adaptation practices are found to be linked with

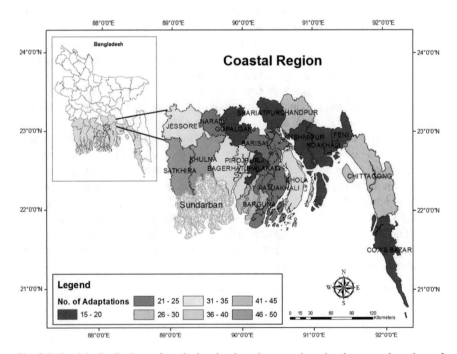

Fig. 2.1 Spatial distribution of agricultural adaptation practices in the coastal region of Bangladesh. *Source* Authors

infrastructural development such as construction of polders for protecting agricultural lands from flood, construction of canals to supply irrigation water, and construction of canals, regulators and sluices to ease drainage congestion of the farm lands. Bangladesh has opted for increasing crop production through large-scale infrastructural development since the mid-1950s. The focus has been on disaster prevention through the construction of physical structures. Since the early 1990s, these structures have been rehabilitated and newly constructed, accommodating potential impacts of climate change. About 25% of the adaptation practices relate to improved crop varieties, such as salinity- and submergence-tolerant rice and non-rice crops, short-duration pulses, and cultivation of cash crops like sunflower, watermelon and sesame. The Bangladesh Rice Research Institute (BRRI) and Bangladesh Institute of Nuclear Agriculture (BINA) have developed salt-tolerant rice varieties such as BRRI dhan41, BRRI dhan47, BRRI dhan53, and BINA dhan7, short-duration rice varieties such as BINA dhan4, and submergence-tolerant rice varieties such as BRRI dhan51 and BRRI dhan52. The Bangladesh Agricultural Research Institute (BARI) has developed different varieties of pulses, oilseeds, fruits and vegetables which are suited to the coastal zone (Ali et al. 2014; Rashid et al. 2014). Among the non-rice crops, BARI Gom25 (wheat), maize, BARI mug6 (mungbean), BARI felon1 (cowpea), BARI til4 (sesame), soybean, Hysun 33 (sunflower), coconut and watermelon are relatively saline-tolerant and cultivated in the Rabi season. About 13% of the practices relate to innovative cultivation techniques like floating bed agriculture, vertical horticulture, pyramid cropping, cultivation of rice alongside vegetables/fish/ducks in a *Sorjan* system, integrated cultivation of rice, fish and vegetable in gher, and integrated farming of rice and fish by creating a ring-microhabitat for fish in a rice field (Ali et al. 2012; Satter/Abedin 2012; Alauddin/Rahman 2013). There is evidence that these innovative cultivation techniques are becoming popular among the coastal people. About 7% of the adaptations have been for the adoption of farm machineries. Other forms of identified adaptations included the creation of irrigation facilities, adoption of agricultural technologies, changes in cropping patterns, livelihood enhancement, and training and capacity building.

Though there can be a number of aims for which adaptations take place, they can be put into three broad categories: implementing action, building capacity, and changing legislation/policy. Whilst other categorisation is possible, these are the ones used for the purpose of this study. The majority of the adaptation practices (about 92%) were found to be for the purpose of implementing actions which reduce hazards, enhance production, and bring direct, tangible benefits to the adaptors. These include introduction of stress-tolerant crop varieties, construction of polders, drainage structures and irrigation canals, integrated farming of crops and fish, and adoption of farm machinery. About 6% of the adaptation practices related to changing legislation/policy to facilitate the adaptation process. Only about 2% of the practices were for building capacity of the coastal people through training and knowledge dissemination, livelihood restoration and rehabilitation, etc.

Adaptations can also be classed into deliberate and non-deliberate adaptations. A deliberate adaptation occurs as a result of real or perceived changes in climate, whereas a non-deliberate adaptation is an action that addresses a non-climatic issue but creates co-benefits. About 95% of the agricultural adaptation practices were identified as deliberate adaptations. These include construction and rehabilitation of polders, drainage structures and irrigation canals, introduction of stress-tolerant crop varieties, and integrated farming techniques. Only some policies and framework related adaptation practices, and some innovative cropping techniques like dyke cropping and floating vegetable gardening, have been identified as non-deliberate adaptations. Due to the nature of the literature review it may have been more difficult to identify non-deliberate adaptations.

2.3.4 Drivers of Adaptation Practices

The majority of agricultural adaptations (about 63%) were found to be reactive in timing, i.e., adaptation occurred following the impacts of climate change. One example of reactive adaptation is the introduction of salt-tolerant crop varieties by various organisations in the coastal areas of Bangladesh where salinity intrusion has been a long-term issue and has caused damage to agricultural activities (Rahman 2012; Titumir/Basak 2012). 37% of the adaptation practices were anticipatory in timing, which indicates that these adaptations had taken place before the climatic impacts became evident, but with an anticipation of changes in the future. For example, the Bangladesh Water Development Board (BWDB) constructed 220 km of irrigation canals and four pumping stations to provide surface water irrigation to 6450 hectares of land in the Chandpur district to boost crop production during the dry season, in anticipation of irrigation water scarcity due to climate change (Huda et al. 1991).

Adaptation practices may evolve in response to a long-term chronic stress or a short-lived sudden shock. It is found that most of the agricultural adaptations (about 76%) have been in response to chronic stresses such as waterlogging and salinity intrusion. The practices to adapt to such situations include construction of drainage infrastructure, irrigation canals and pumping stations, introduction of saline and submergence-tolerant crop varieties, rainwater harvesting for irrigation, and alternative livelihood practices like homestead vegetable gardening and integrated cropping and livestock rearing. About 24% of the adaptation practices have emerged in response to sudden shocks such as major floods, cyclones and storm surges. The adaptation practices that have evolved to adapt to such shocks include construction of embankments, livelihood restoration and rehabilitation, and practices strengthening household capacity by taking multiple livelihood options, training activities, etc.

2.3.5 Providers and Beneficiaries of the Adaptations

In the coastal area of Bangladesh, the government has been the major provider of agricultural adaptations. BWDB, BRRI, BARI and the *Comprehensive Disaster Management Program* (CDMP) are the principal government organisations providing about 78% of the total adaptations, while various national and international NGOs have provided about 21% of the adaptations. Among the NGOs, the Bangladesh Rural Advancement Committee, Care Bangladesh, Oxfam International, Practical Action and Action Aid provided the majority of the adaptations. The role of the private sector as an identified adaptation provider has been insignificant (1%). Though it is generally known that the private sector is involved in commercial production of hybrid seeds, particularly for vegetables, and is a dominant actor in agricultural value chain, its role as an adaptation provider is not studied and documented in adaptation literature.

The local people of the targeted area are reported to be the beneficiaries for the majority of the agricultural adaptations (about 92%). Among the direct beneficiaries are community people, farmers, fishermen, disaster affected people, and vulnerable women and children. Since the adaptations were agriculture related, the majority of the beneficiaries were the farmers. It is the particular community as a whole, rather than individuals within the community, who benefited through many of the adaptations. This indicates that agricultural adaptation has a good distribution of benefits and has a better equity outcome. Commercial farmers with large landholdings experienced high benefits from large-scale adaptations such as physical structural adaptations and agricultural mechanisation, whereas subsistence farmers with small landholdings benefitted more through small-scale adaptations such as integrated farming and homestead vegetable cultivation. Among different agricultural actors, the farmers and wage-labourers are the major victims of climate change and variability. They mostly bear with the losses associated with hazards. Other actors such as traders, processors and input sellers usually have alternative means and strategies to cope with climatic shocks and stressors.

2.3.6 Barriers to Adaptation Practices

A number of barriers were reported in relation to the adaptation practices in the coastal zone of Bangladesh. Common barriers identified include a lack of capital, lack of access to resources, lack of knowledge and information, lack of coordination, and lack of institutional capacity. About 45% of the adaptation practices have faced at least one barrier.

Since many of the adaptations require a large investment, capital cost is a dominant limiting factor. Natural resources, such as land, are required for the implementation of some of the adaptations, but proved to be difficult for implementing organisations to acquire. Lack of knowledge and information on quality inputs such as seeds, fertilizers, and pesticides, and lack of capacity and coordination among the responsible organisations were also identified as barriers. The prevailing unfavourable agro-environment itself, such as excessive soil moisture at the beginning of Rabi season, lack of freshwater for irrigation, and higher than tolerable limits of soil salinity, were also among the barriers. More importantly, the very organic structure and decision-making processes of some of the government organisations are not conducive to community participation. Overcoming these barriers will be important for the sustainability of the adaptation practices.

Though there is evidence of sustainability of some adaptation practices through replication, community uptake, gain in popularity, transformation of land use practices, enrichment of environmental quality, and increased access of marginalised and women groups to assets (Abedin/Shaw 2001; Nandy et al. 2013; Rahman/Islam 2013), sustainability issues were not reported for many of the adaptation practices. Some stress-tolerant crop varieties, vegetable farming on floating mats, cultivation of some non-rice crops such as soybeans and sunflowers, embankment cropping, integrated farming of rice, fish and vegetables, vertical land management with fish cultivation in the lower part and vegetation (horticulture/forestation) in the upper part, etc., were found to be among the sustainable practices.

2.3.7 Gender Issues

Though it is generally understood that there are some gender-specific issues in almost all the adaptations, these issues are reflected in only 27% of the agricultural adaptation practices. This is reflected mainly in terms of the involvement of women in post-harvest activities like thrashing, drying, sorting, grading, storing, etc. The cultivation of sunflowers and watermelons in the coastal areas are mainly done by women, as it is less labour-intensive than rice (Saha 2016). Also, adaptations involving livelihood restoration and rehabilitation are gender sensitive, as they include technical training and institutional support for farming, fishing, cattle and poultry rearing, nursery management, and small agro-entrepreneurship, to the targeted vulnerable groups such as women-headed households, widows, divorced, small farmers, people with disabilities, vulnerable households, landless people, youth, and school children. The majority of the direct beneficiaries of such adaptations are the poor women (Rahman 2013). Alternative income generation through homestead vegetable production and small cottage industries, such as weaving reed mats, were noted as gender sensitivity inclusive and considering the welfare of women. In 24% of the adaptation practices, there has been a deliberate attempt to make the practices gender appropriate. Nevertheless, it is a fact that gender-specific issues are not reflected in about 73% of the agricultural adaptation practices in the coastal area of Bangladesh, though there is a strong commitment in agriculture policies and plans (GED 2011; MoA 2013a). Bangladesh has prepared a separate gender action plan to ensure gender equality in climate related policies, strategies and interventions (MoEF 2013). Involvement of women in agriculture through the use of alternative technologies such as bio-fertilisers, climate-resilient cropping, etc., has been suggested in the plan.

2.3.8   Policies and Plans

Adaptation, low carbon development, mitigation, technology transfer, disaster risk reduction, and the mobilisation and international provision of adequate finance have been identified as means to manage climate change impacts in Bangladesh (MoEF 2009a). Of these, the Government prioritises adaptation and disaster risk reduction. A number of policies, strategies and plans are already in place to tackle the adverse impacts of climate change on different socioeconomic sectors including the agriculture sector. The *Bangladesh Climate Change Strategy and Action Plan* (BCCSAP) (MoEF 2009a) is built on six pillars and comprises 44 programs of different timeframes. Five of the six pillars and 24 of the 36 programmes are linked to the agriculture sector. The *National Adaptation Program of Action* (NAPA) (MoEF 2009b) identifies 38 adaptation measures in eight thematic areas. It highlights the measures to be taken by the different government line ministries and departments. However, it fails to integrate the capacities, knowledge bases, and resources of NGOs in tackling climate change impacts. Climate change has received some attention as a crop sector challenge in the Government's 6th Five Year Plan (GED 2011). In addition, encouraging agricultural research on adaptation to climate change is among the core objectives for the crop sector in this plan. Among the sectoral policies and plans, the National Water Policy (MoWR 1999) has a separate sub-section on 'water and agriculture', though it does not address climate change issues. Climate change has also appeared as a passing reference in a few places in the recently declared National Agriculture Policy (MoA 2013a). The Master Plan for Agricultural Development in the Southern Region of Bangladesh (MoA 2013b) mentions climate change and sea level rise as a challenge, and suggests redesign and new construction of polders as climate-resilient structures. It also considers 'response to climate change threat' as one of the four criteria in determining the urgency of a programme required for prioritising investment. Thus, it appears that a few policies and plans are already in place in Bangladesh to tackle climate change issues, though some of these, particularly the sectoral policies, need updating with respect to climate change. The policies, particularly BCCSAP and NAPA, are well cited in adaptation documents in Bangladesh and are often referred to in climate change related programmes and projects. More importantly, adaptation activities have been implemented across all thematic areas of BCCSAP. Also, the Government of Bangladesh has proactively constituted a Climate Change Trust Fund and has been allocating funds from its own revenues since the 2009–2010 fiscal year to address climate change vulnerabilities.

2.4   Further Discussions and Future Directions

Bangladesh is universally acknowledged as one of the most vulnerable countries to climate change due to a number of physical and socioeconomic factors. Being heavily dependent on seasonal climatic and hydrologic conditions, crop agriculture is particularly vulnerable to climate change impacts. Despite this high vulnerability, the country's crop sector has made commendable progress in maintaining food production (FAO et al. 2015). This progress can be put down to better input uses (seeds, irrigation, etc.), institutional support through extension services, credit facilities and capacity building, agricultural research innovation, and investment in disaster risk reduction measures (flood protection embankments, drainage structures, etc.). The success is reflected in the wide range of adaptation practices that Bangladesh has embraced to adapt to adverse climatic and environmental conditions. However, there are some gaps between policies, potential adaptations and the adaptations in practice, particularly in the coastal area.

Policies thus far favour improved crop varieties as an adaptation measure. Hundreds of crop varieties, many of which can tolerate stresses and resist pests and diseases, have been created in the country. However, there are only a few varieties introduced in practice and the scale of adoption is not that high. The reason could be that the productivity of the already developed cultivars is not promising against severe stresses (MoA 2013a, b). For saline-tolerant rice, lodging tendency is reported to be a problem for BRRI dhan41 and BRRI dhan47 (Rashid et al. 2014). Lack of awareness among the farmers on the advantages of this technology, as a result of inadequate extension services, is another reason for its poor adoption (Mainuddin et al. 2011). We also found, through two recent field visits to the southern part of the country, that the saline-tolerant rice varieties have not become popular among the farmers. Farmers are still cultivating local improved varieties. Clearly, there is a gap between research and farmers' uptake, and this gap needs to be narrowed by strengthening agricultural extension services and research, and improving coordination among and capacity of the service providers.

Some of the potential adaptations identified in the literature, such as improving irrigation efficiency, solar-powered irrigation, crop diversification, conjunctive use of surface and groundwater, and development of agro-businesses, have not yet been reflected in the policies and practices of the coastal area. Initiatives could be undertaken through the Department of Agricultural Extension, Department of Agricultural Marketing, or NGOs to demonstrate such adaptations in the field, and to sensitise, raise awareness among, and motivate the farmers to engage with such adaptations. There is a lack of knowledge and information among farmers and extension workers about input uses during a hazard. This may be because a low proportion of adaptation practices involve training and capacity development. Whilst there is a good presence of government organisations and NGOs as adaptation providers, the private sector is almost absent – this could be an area for future engagement and improvement. Since farmers can be considered as micro-entrepreneurs, and they are one of the main vulnerable groups, this is an area of considerable importance. Moreover, initiatives that do not consider uptake by farmers and the private sector might not be sustainable in the long-term.

It is found that regular and periodic maintenance is often lacking in the case of physical structures. This leads to structures in a dilapidated condition that collapse in the event of a moderate shock. Also, it usually takes a long time to restore the structures once they are damaged in a disaster. Lack of funding, slow decision-making, and lack of personal and organisational integrity and account-ability are responsible for this. The work culture in some organisations is not founded on a bottom-up approach, which may be a barrier to taking up community based adaptations (also see Wright et al. 2014). In addition, adaptation practices need to internalize gender issues, as only a small share of current adaptation practices specifically consider them. An integrated management approach to land, water, and socio-ecological systems is needed to stop ongoing environmental deterioration and to ensure food security. Land use planning and regulation, and integrated production systems, can help maintain the balance. There are currently few studies that take an integrated approach to adaptation in Bangladesh, or syn-thesize the literature as this article has done. Thus, it is recommended that more studies based on holistic approaches are needed.

Acknowledgement This work is carried out under the *Deltas, Vulnerability and Climate Change: Migration and Adaptation* (DECCMA)' project under the *Collaborative Adaptation Research Initiative in Africa and Asia* (CARIAA) program with financial support from the UK Government's *Department for International Development* (DFID) and the *International Development Research Centre* (IDRC), Canada. The views expressed in this work are those of the authors and do not necessarily represent those of DFID and IDRC or its Boards of Governors. The comments and suggestions provided by Dr. Helena Wright of the Center for Environmental Policy, Imperial College London, UK and the two anonymous reviewers helped improve the quality of this manuscript and are gratefully acknowledged.

References

Abedin MA, Shaw R (2001) Agriculture adaptation in coastal zone of Bangladesh. In: Shaw R, Mallick F, Islam A (eds) Climate change adaptation actions in Bangladesh. Springer, Japan, p 207–225

Ahmed AU (2006) Bangladesh climate change impacts and vulnerability: a synthesis. Climate Change Cell, Department of Environment, Comprehensive Disaster Management Program, Dhaka

Ahmed AU (2010) Reducing vulnerability to climate change: the pioneering example of community-based adaptation. Center for Global Change, Dhaka

Alauddin SM, Rahman KF (2013) Vulnerability to climate change and adaptation practices in Bangladesh. J SUB 4(2): 25–42

Ali A (1999) Climate change impacts and adaptation assessment in Bangladesh. Clim Res 12: 109–116

Ali MO, Zuberi MI, Sarker A (2012) Lentil relay cropping in the rice-based cropping system: an innovative technology for lentil production, sustainability and nutritional security in changing climate of Bangladesh. J Food Sci Eng 2: 528

Asaduzzaman M, Ringler C, Thurlow J, Alam S (2010) Investing in crop agriculture in Bangladesh for higher growth and productivity, and adaptation to climate change. Bangladesh Food Security Investment Forum, Dhaka

Asia Foundation (2012) A situation analysis of climate change adaptation initiatives in Bangladesh, Dhaka

BBS (2011) 2011 yearbook of agricultural statistics of Bangladesh. Bangladesh Bureau of Statistics, Dhaka

BBS (2013) Statistical yearbook of Bangladesh 2012. Bangladesh Bureau of Statistics, Dhaka

Brammer H, Asaduzzaman M, Sultana P (1993) Effects of climate and sea-level changes on the natural resources of Bangladesh. Bangladesh Unnayan Parishad, Dhaka

Dev PK (2013) Water logging through soil-less agriculture as a climate resilient adaptation option. In: Filho WL (ed) Climate change and disaster risk management. Springer, p 681–692

Ericksen NJ, Ahmad QK, Chowdhury AR (1993) Socio-economic implications of climate change for Bangladesh. Bangladesh Unnayan Parishad, Dhaka

FAO, IFAD, WFP (2015) The state of food insecurity in the world. Food and Agriculture Organization of the United Nations, International Fund for Agricultural Development and World Food Program

Faruque HSM, Ali ML (2005) Adaptation to climate change for managing water sector. In: Mirza MQ, Ahmad MQ (eds) Climate change and water resources in south Asia. Taylor and Francis, p 231–254

GED (2011) 6th Five Year Plan FY 2011-FY 2015: accelerating growth and reducing poverty. General Economics Division, Planning Commission, Dhaka

Huda ATMS, Huq MDE, Bhuyian AI (1991) The Bangladesh Meghna-Dhangoda irrigation project. Public Admin Develop 11: 205–218

IPCC (2014) Climate change 2014: synthesis report, annex II: glossary. Intergovernmental Panel on Climate Change, p 117–130

Karim Z (2011) Assessment of investment and financial flows to adapt to the climate change effects in the agriculture sector. Government of the People's Republic of Bangladesh, Dhaka

Karim Z, Hussain SG, Ahmed AU (1998) Climate change vulnerability of crop agriculture. In: Huq S, Karim Z, Asaduzzaman M, Mahtab F (eds) Vulnerability and adaptation to climate change for Bangladesh. Kluwer Academic Publishers, p 39–54

MoA (2013a) National agriculture policy 2013. Ministry of Agriculture, Dhaka

MoA (2013b) Master plan for agricultural development in the southern region of Bangladesh. Ministry of Agriculture, Dhaka

MoEF (2009a) Bangladesh climate change strategy and action plan 2009. Ministry of Environment and Forests, Dhaka

MoEF (2009b) National adaptation program of action (NAPA). Ministry of Environment and Forests, Dhaka

MoEF (2013) Bangladesh climate change and gender action plan (ccGAP: Bangladesh). Ministry of Environment and Forests, Dhaka

Mondal MS, Islam AKMS, Modhu MK (2012) Spatial and temporal distribution of temperature, rainfall, sunshine and humidity in context of crop agriculture. Comprehensive Disaster Management Program, Ministry of Food and Disaster Management, Dhaka

Mondal MS, Saleh AFM, Akanda MAR, Biswas SK, Moslehuddin AZM, Zaman S, Lazar AN, Clarke D (2015) Simulating yield response of rice to salinity stress using the AquaCrop model. Environ Sci Process Impact 17(6): 1118–1126

MoWR (1999) National water policy. Ministry of Water Resources, Government of the People's Republic of Bangladesh, Dhaka

Oxfam International (2009) Climate change adaptation practices in thirty agroecological zones (AEZs) of Bangladesh, Final Report. Jahangirnagar University, Dhaka

Rabbani G, Rahman A, Ishtiaque JS, Khan ZM (2015) Climate change and food security in vulnerable coastal zones of Bangladesh. In: Habiba U, Abedin MA, Hassan AWR, Shaw R (eds) Food security and risk reduction in Bangladesh. Springer, Japan, p 173–185

Rahman MM (2012) Enhancement of resilience of coastal community in Bangladesh through crop diversification in adaptation to climate change impacts. MS thesis, Postgraduate Programs in Disaster Management, BRAC University, Dhaka

Rashid MH, Islam MR, Saleque MA, Muttaleb MA, Mondal S, Ali MH, Sattar SA, Abedin MZ (2014) Farming systems options to adapt with climate change in south western Bangladesh. Res Crop Ecophysiol 9(2): 70–80

Rawlani AK, Sovacool BK (2011) Building responsiveness to climate change through community based adaptation in Bangladesh. Mitig Adapt Strateg Glob Change 16: 845–863

Saha D (2016) Prospect of Rabi crops in south-west coastal area under climate change scenarios. MSc (WRD) thesis, Institute of Water and Flood Management, Bangladesh University of Engineering and Technology, Dhaka

Saleh AJM, Mondal MS (2009) Irrigation intervention as a tool of socio-economic advancement: a case study of the Kawraid River Rubber Dam Project, Gazipur, Bangladesh. In: Proc Second Int Conf on Water Flood Manage, Institute of Water and Flood Management, Dhaka, 15–17 March, p 641–650

Sattar SA, Abedin MZ (2012) Options for coastal farmers of Bangladesh adapting to impacts of climate change. In: International Conference on Environment, Agriculture and Food Sciences, Thailand, p 17–21

Sterret C (2011) Review of climate change adaptation practices in South Asia, Oxfam Research Report, Oxfam

Titumir RAM, Basak JK (2012) Effects of climate change on crop production and climate adaptive techniques for agriculture in Bangladesh. Soc Sci Rev 29(1): 215–232

Tompkins EL, Adger WN, Boyd E, Nicholson-Cole S, Weatherhead K, Arnell N (2010) Observed adaptation to climate change: UK evidence of transition to a well-adapting society. Glob Environ Chang 20: 627–635

Tompkins EL, Boyd E, Nicholson-Cole S, Weatherhead K, Arnell N, Adger WN (2009) An inventory of adaptation to climate change in the UK: challenges and findings. Tyndall Center for Climate Change Research

World Bank (2000) Bangladesh: climate change and sustainable development. The World Bank, Rural Development Unit, South Asia Region

World Bank (2011) The cost of adapting to extreme weather events in a changing climate. Bangladesh development series, Paper No. 28, Dhaka

Wright H, Vermeulen S, Laganda G, Olupot M, Ampaire E, Jat ML (2014) Farmers, food and climate change: ensuring community-based adaptation is mainstreamed into agricultural programs. Clim Dev 6(4): 318–328

Younus MA, Bedford RD, Morad M (2005) Not so high and dry: patterns of 'autonomous adjustment' to major flooding events in Bangladesh. Geography 90(2): 112–120

Chapter 3
Climate Change-Induced Loss and Damage of Freshwater Resources in Bangladesh

Nandan Mukherjee, John S. Rowan, Roufa Khanum, Ainun Nishat and Sajidur Rahman

Abstract Climate change loss and damage is evident in hydrological perturbations among river systems in Bangladesh. Significant disruptions include changes in the intensity, frequency, and seasonality of peak and low flow characteristics. Over the last few decades, water-related disasters conveyed through the river systems have caused increased economic damage of assets and infrastructure. Other impacts include the loss of fish spawning grounds and reduced agricultural production due to changes in the hydrological regime. This chapter discusses a broad range of generalised approaches to address water-related disasters and changes in hydrological characteristics.

Keywords Climate change · Loss and damage · Hydrological alteration
Riverine ecosystem

3.1 Introduction

Climate change will induce loss and damage especially through its impact on freshwater resources. Evidence worldwide strongly suggests that climate change will perturb hydrological systems by altering the frequency, intensity, and spatial and seasonal distribution of precipitation (Bates et al. 2008). For example, greater atmospheric moisture content over the warmer Indian Ocean will generally intensify South Asian summer monsoon rainfall and generate stronger extreme weather

Nandan Mukherjee, University of Dundee, Dundee, Scotland, Corresponding Author, e-mail: n.mukherjee@dundee.ac.uk.

John S. Rowan, University of Dundee, Dundee, Scotland.

Roufa Khanum, Centre for Climate Change and Environmental Research, BRAC University, Dhaka, Bangladesh.

Ainun Nishat, Centre for Climate Change and Environmental Research, BRAC University, Dhaka, Bangladesh.

Sajidur Rahman, Centre for Climate Change and Environmental Research, BRAC University, Dhaka, Bangladesh.

© Springer Nature Switzerland AG 2019
S. Huq et al. (eds.), *Confronting Climate Change in Bangladesh*,
The Anthropocene: Politik—Economics—Society—Science 28,
https://doi.org/10.1007/978-3-030-05237-9_3

events, but will also lead to less predictable changes to intraseasonal and interannual variability (Turner/Annamalai 2012). The *Intergovernmental Panel on Climate Change* (IPCC) reports that freshwater availability likely will be severely affected and the potential for conventional adaptation based approaches will be limited if the temperature increases by more than 2 °C (Cisneros et al. 2014). Bangladesh's population of nearly 160 million (BBS 2016) relies heavily upon its hydrological systems and is therefore vulnerable to events which will be exacerbated by climate change, e.g., floods, drought, storm surges, sea level rise, and salinity intrusion. Moreover, unplanned development within river basins and freshwater withdrawal from transboundary rivers by the upper riparian countries will likely intensify climate change-induced loss and damage from salinity intrusion and sea-level rise (Mirza 1998; Rahman et al. 2011).

A range of adaptation and mitigation approaches is outlined in the *Bangladesh Climate Change Strategy and Action Plan* (BCCSAP) (MoEF 2009). However, these proposals need to be geographically specific with regards to financing and potential climate impacts. Depending on local hydrological and socioeconomic characteristics, infrastructural adaptation options—e.g., flood control and drainage structures—may be constrained in preventing significant loss and damage (Kudzewicz et al. 2014). Persistent development deficits may also limit adaptive capacity among affected populations.

This chapter aims to assess the climate change-induced losses and damages pertinent to freshwater systems in Bangladesh and outline broad-ranged approaches to address them. Section 3.2 presents an overview of hydrological conditions and loss and damage from water-related disasters in Bangladesh. Section 3.2.2 reviews the state of current water management systems addressing loss and damage. Section 3.2.5 explores potentially sustainable options for managing residual loss and damage.

3.2 Hydrological Loss and Damage

3.2.1 Climate Change and Variability in Bangladesh

Bangladesh is divided into seven hydrological regions, with most surface water flowing through drainage systems comprised of three transboundary rivers: the *Ganges, the Brahmaputra, and the Meghna* (GBM) (WARPO 2001). Around 700 rivers with tributaries flow through the deltaic floodplain into the Bay of Bengal (Khalil 1990). Glacier contribution is relatively insignificant to the total flow contribution, with rainfall runoff forming more than 90% of the discharge (Jain 2008). Therefore, flow extremes such as floods and droughts are highly sensitive to rainfall conditions. Bangladesh experiences three types of freshwater flooding (Ahmed 2005): (1) flash floods triggered by overflowing of hilly rivers in the East and North during the spring and autumn; (2) rainfall-induced floods resulting from heavy precipitation and drainage congestion; and (3) monsoon riverine floods in the floodplains caused by excess flow in the GBM basins.

Perturbations to the hydrological cycle resulting from climate change and variability affect the intensity and frequency of water-related hazards (Fig. 3.1). For example, a rise in atmospheric temperature can reduce soil moisture content and increase the risk of agricultural drought, and changes in precipitation volume and timing affect flood incidence and exposure. In the latter half of the 20th century, there was an observable increase in the incidence of warm temperature extremes, though changes in precipitation extremes have been less apparent (Klein Tank et al. 2006). Other analyses show that there has been little change to the June–September seasonal mean South Asian summer monsoon rainfall since the mid-twentieth century (Turner/Annamalai 2012). Thus, consistently significant climate-induced alterations to peak flow and discharge regimes have yet to be recorded (Mirza et al. 2001; Gain et al. 2013). However, modelled climate change scenarios suggest that in the future flooding will increase in geographical extent, intensity, and frequency,

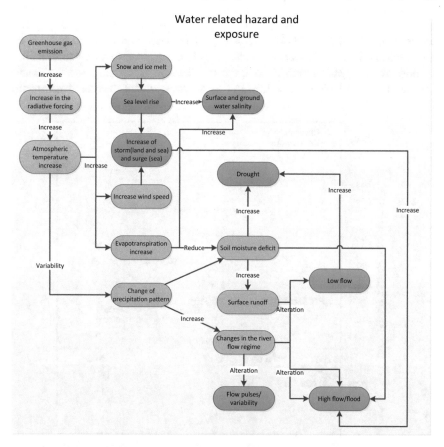

Fig. 3.1 Climate change impact on water-related hazards. *Source* Adopted and modified from Mirza/Ahmad (2005: 10)

primarily due to Brahmaputra and Meghna peak discharges (Mirza et al. 2003; Gain et al. 2013).

In northern and northwestern Bangladesh, low flow conditions also frequently occur due to a lack of precipitation during the dry season (i.e., winter to pre-monsoon) (Shahid/Behrawan 2008). While pre-monsoon drought hinders the minor winter cropping season, delayed or reduced monsoons can significantly impact primary rice and inland fisheries production (Agrawala et al. 2003; Faruque/Ali 2005). More extreme wet-to-dry season flow ratios will also affect sediment transport and deposition, resulting in channel sedimentation and riverbank erosion.

3.2.2 Direct, Tangible, and Economic Loss and Damage

Accounting for loss and damage from water-related hazards needs to cover both tangible and intangible impacts. Tangible, direct loss and damage include loss of human life, emergency assistance to affected people, and economic damages to property and infrastructure. The frequency of recorded water-related hazards has increased significantly in recent decades, largely due to increased human settlement in flood-prone areas as well as to improved damage monitoring and assessment (Mirza et al. 2001) (Fig. 3.2). EMDAT is an international disaster database that

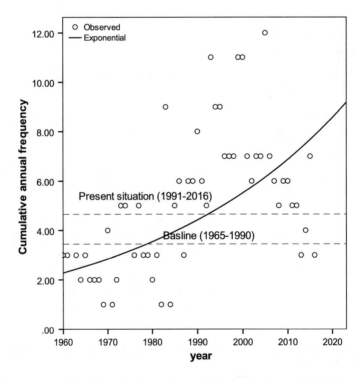

Fig. 3.2 Frequency of hydrological disasters from 1960 to 2016. *Source* EMDAT data

systematically reports information regarding historical disaster incidences (Guha-Sapir et al. 2017), using several qualifying filters. Disaster events are included when at least ten mortalities occurred, at least 100 people were affected, there was a declaration of a state of emergency, or a call for international assistance by the state was announced.

There were 58 cases of water-related disasters reported in Bangladesh between 1965 and 1990, a count which nearly doubled to 113 between 1991 to 2016. Storm surge and floods comprise 85% of all the hazard incidences. On average, the hazard frequency has been increasing at the rate of approximately 7% per year.

In contrast, the loss estimates, i.e., deaths and the total affected human population, has declined in recent years. The death toll has declined by 57%, from a mean annual incidence of 375 between 1965 and 1990 to 159 between 1990 and 2015. The total affected population has also declined between these two periods, decreasing from 6.8 million to 3.4 million per year. Over the last few decades, national efforts in disaster risk reduction and risk management, such as protective infrastructure and early warning systems, may be contributing to vulnerability mitigation.

Normalised economic damage estimates (Neumayer/Barthel 2011) increased nearly 65%, from a cumulative estimate of US$68 million between 1965 and 1990 to US$324 million between 1990 to 2015, growing at an average rate of US$9 million per year (Fig. 3.3). However, this trend is not solely attributable to climate

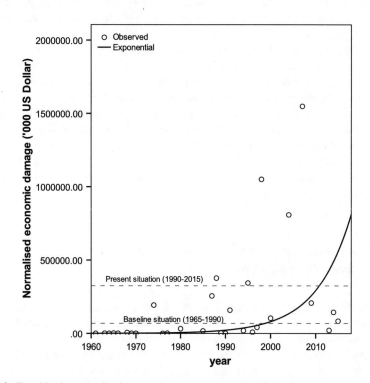

Fig. 3.3 Trend in the normalised economic damage. *Source* EMDAT data

change and variability, since economic development in disaster prone areas has increased the quantity of vulnerable property and infrastructure.

3.2.3 Indirect, Intangible, and Non-economic Loss and Damage

Indirect dimensions of loss and damage may be more pervasive and long-term, such as hydrological alteration of river flows which negatively affect fish spawning and fishery-based livelihoods, as well as modified flooding regimes which impact agriculture in flood-prone areas.

For example, the Halda River, located in the southeastern hilly region of Bangladesh, is a critical spawning ground for three major carp species: *Catla catla*, *Labeo rohita* and *Cirrhinus mrigala*. The spawning biology for these important fishery species are sensitive to ecological characteristics which are determined by meteorological and flow conditions (Tsai et al. 1981). Spawning occurs following a sudden rise in water level due to thunderstorms during monsoon floods from April to June, and may be triggered by the increased current, turbulence, or up-welling. Therefore, perturbations in the timing and intensity of high flow conditions increase the potential loss of spawning by these migratory fishes. Another example of indirect loss and damage occurs in the haor basin of Bangladesh. Located in the northeast, the haor basin is a mosaic of wetland habitats—including rivers, irrigation canals, seasonal cultivated floodplains, lakes and freshwater swamps—that resembles a bowl-shaped depression (Nishat et al. 2002). This internationally important wetland ecosystem is home to numerous floral and faunal species and migratory birds, and is also one of the largest dry season (boro) rice producing areas of Bangladesh (Uddin et al. 2013). Flash flooding is characteristic of haor regional rivers, which often impedes the boro harvest during the month of May. During the 2004 flood, more than two thirds of the boro production was lost due to an early flash flood event that coincided with the harvest season (CEGIS 2012). Increased frequency and intensity of such events due to climate change would therefore have a deleterious impact on the boro rice harvest.

3.2.4 Approaches to Reduce Loss and Damage from Hydrological Disasters

Flood control, drainage, and irrigation (FCDI) have been the most common forms of structural intervention for water resource development in Bangladesh. These include river embankments, dam construction, reforestation, drainage channels, and pump drainage (Faruque/Ali 2005). Such interventions are the principle adaptation options for minimising the loss and damage from monsoon flooding in the north-west and north central region, tidal flooding in the northeast region, coastal flooding and storm surge in the southern coastal region, and water scarcity and irrigation demand mitigation in the northwest and southwest region.

Typical FCDI strategies are organized around three annual rice cropping seasons: (1) Kharif-I (mid-April to mid-July); Kharif-II (mid-July to mid-November) and Rabi (mid-November to April). The strategies are designed to protect crops against early river floods during Kharif-I; expand the cultivated area through flood exclusion during Kharif-II; and retain water in the system to reduce drought risk during the Rabi (Faruque/Ali 2005). The main objective is to reduce losses due to flooding or drought, in order to encourage farmers to invest in the greater inputs required by high yielding food crops, resulting in greater agricultural productivity. However, flood control structures which inhibit fish migration and reduce annual floodplain inundation may reduce the productivity of freshwater capture fisheries, while allowing culture fisheries to expand within protected zones.

The *Bangladesh Water Development Board* (BWDB), as the principle implementing authority under the Ministry of Water Resources, has completed roughly 700 FCDI projects since the early 1950s. The BWDB estimates that the FCDI projects have cost about US$3 billion (unadjusted) since their inception (Nishat et al. 2011). More recently, the United Nations Development Programme has estimated that new water resources development projects have cost US$1.3 billion between 1999 and 2011. In terms of the total investment, FCDI projects are dominated by flood control, drainage, and irrigation projects (24%); town protection projects (22%); river bank protection projects (16%); and irrigation and command area development (10%) (Fig. 3.4).

The *National Water Management Plan* (NWMP) (WARPO 2001), prepared in line with the National Water Policy (MoWR 1999), is the latest long-term strategy for water resources. Although the NWMP did not directly address climate change adaptation, it outlined 84 project portfolios regarding water and river resources, emphasizing year-round water management, stakeholder engagement, and social and environmental needs (Faruque/Ali 2005). Subsequently, the *Ministry of Environment and Forest* (MoEF) enacted the first *National Adaptation Plan of Action* (NAPA) (MoEF 2005), which reiterated the continued importance of building FCDI infrastructure as an adaptation option. More recently, the *Bangladesh Climate Change Strategy and Action Plan* (BCCSAP) (MoEF 2008) revisited the NAPA propositions and recommended a total of 44 projects under six thematic areas including infrastructure development for combating natural disasters. BCCSAP also proposed the retrofitting of existing FCDI infrastructure as well as the building of additional infrastructure-based protection for exposed coastal zones and river flood plains.

Despite noticeable progress in water resources development, some initiatives may be maladaptive despite claimed benefits in the short term. For example, flood or coastal protection embankments and drainage control structures reduce vulnerability by preventing floods or storm surges from reaching the settlements. However, in most cases, the intended benefit occurs only for a short period. In the longer term, flood embankments restrict the sediment inflow to the flood plain, which ultimately reduces the nutrient availability of the topsoil and raises the

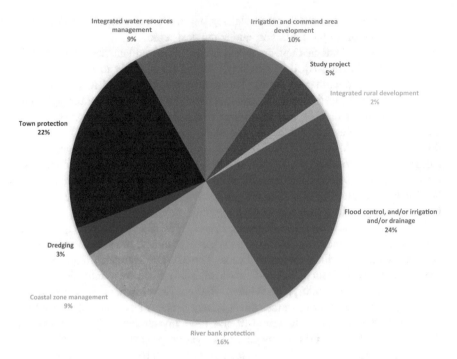

Fig. 3.4 Cumulative investment proportion in different categories of water sector development projects (1999–2011). *Source* Graph is drawn using the data from annual development program, Ministry of Finance, Gob, adopted from Nishat et al. (2011)

adjacent river bed, causing an imbalance in the sediment budget of the river system (Brouwer et al. 2007). Siltation and poor maintenance of drainage channels further exacerbates congestion and waterlogging (Faruque/Ali 2005).

Flood embankments constructed upstream increase the risk of flooding and geomorphological instability (i.e., erosion or accretion) in the downstream sections (see Jauhari/Gadhalay 2011). Earthen embankments also often create a false sense of security among the inhabitants living inside the embanked area, resulting in a greater hazard when the structures are breached or overtopped by megafloods or storm surges. Permanent waterlogging within poldered areas in Southwestern Bangladesh (Brammer 1983; Nowreen et al. 2014) and loss and damage from mass embankment failure during the Cyclone Sidr (Islam et al. 2010) are documented examples of maladaptive consequences.

Additionally, utilising excessive groundwater or pumping water from rivers are common adaptation options for providing irrigation water for agriculture as rainfall becomes less predictable. However, these practices have long-term impacts on groundwater availability, increasing the risk of land subsidence and compromising natural flows in the river systems (Safiuddin/Karim 2001; Hoque et al. 2007).

Therefore, conventional water management interventions have limits which might cause further loss and damage to the environment and society. Instead, water management interventions should do the following:

1. Reduce the exposure of the vulnerable population, assets and infrastructure efficiently and without invoking a false sense of security. Instead of relying on conventional large scale FCDI projects, ecosystem based adaptation options could be explored (see Chap. 5 of this volume).
2. Proactively plan adaptation options so that the residual loss and damage to society and the environment is minimised. The conventional design of large water management infrastructure mostly relies on historical inundation regimes. However, given the unpredictability of climate change outcomes, uncertainty and the limits of adaptation and mitigation must be considered and incorporated into adaptive water management principles (see Box 3.1).
3. Maximise the potential for complementary or dual benefits from climate adaptation and greenhouse gas mitigation. For example, a coastal green belt could both provide coastal protection and while sequestering carbon (Chow 2017).
4. Manage each river at the basin scale as a single, living system, avoiding country-wide interventions. Calculation of the environmental flow requirements of the river system is long due in river basins around the globe. Recently, New Zealand's parliamentary approval of a bill recognising the Whanganui river as having the same legal rights as human beings—as well as the Ganges and Yamuna rivers having been declared living entities by the government in Uttarakhand, India—are bold examples of such propositions.

3.2.5   Reducing Loss and Damage Through Managing Risk

Risk management efforts, including adaptation, mitigation and disaster risk reduction, are contingent upon perceptions of risks which are acceptable, tolerable, or intolerable depending on the frequency of occurrences and the scale of the consequences (Klein et al. 2014). Acceptable risks are those so minor that action is considered unnecessary, whereas tolerable risks potentially can be kept at reasonable levels through adaptive management efforts. Intolerable risks, however, can threaten human welfare and ecological sustainability due to the lack of practicable or affordable adaptation options. Uncertainty magnifies risk, particularly when climatic events change in terms of intensity, frequency, and spatial and temporal variation. Due to the limits of mitigation and adaptation activities, residual loss and damage from climate change will occur. If loss and damage are considered intolerable, they can either be accepted or avoided via dramatic adaptive transformation of the social or natural system (Klein et al. 2014).

Risk reduction approaches are intended to manage loss and damage of assets, infrastructure, and human life. Risk reduction methods can be divided into two categories, structural and non-structural. Structural approaches mostly reduce risk by altering the exposure dimensions, e.g., reducing the population's exposure to risk from living in the flood plain by constructing new defences or raising the plinth or base of the critical infrastructure above the flood level. These also include retrofitting the height of the flood embankment and drainage infrastructure. However, structural measures may trigger further loss and damage to the environment and ecosystem. Therefore, risk retention and risk pooling measures are necessary to complement risk reduction activities, given the limits of infrastructural adaptation options.

Non-structural risk reduction approaches include early warning systems and community-based disaster management initiatives. Non-structural options mainly address vulnerability issues, either by reducing the sensitivity of the exposed element or by enhancing the adaptive capacity of the same. In Bangladesh, although the Flood Forecasting and Warning Center can forecast flood levels at some locations 48 hours in advance, public awareness systems are limited to print and electronic media (Faruque/Ali 2005; Mirza/Burton 2005). Other non-structural options include zoning, hazard preparedness, shelters, emergency services, and improved public communications and education. However, these activities are not widespread in Bangladesh (Mirza/Burton 2005).

Risk retention measures aim to enhance the capacity of the exposed population to absorb the shocks of loss and damage. The principle focus is on improving adaptive capacity through pre- and post-disaster emergency response, launching social safety net programmes, and also by introducing microfinance tools and instruments. This particular approach works well when risk reduction strategies fail due to the unprecedented scale of natural hazards. Risk pooling options are intended to shift economic risks from an individual or organisation to an insurer (UNFCCC 2012). This approach works well for managing economic loss and damage that cannot be prevented through risk reduction or retention based methods. Microfinance instruments like micro insurance or index-based insurance for agriculture are common risk transfer options.

Most kinds of structural adaptation options related to water resources management may have significant residual environmental and social impacts even when supported by a robust environmental management framework. Intangible and residual loss and damage may occur because of the unprecedented nature of climate change induced hydro-meteorological hazards, which may overwhelm efforts to reduce vulnerability. Ecosystem-based options usually have much less residual impact than traditional structural adaptation strategies. A single, myopic approach is not sufficient to address the broad horizon of loss and damage.

3.3 Conclusion and Recommendations

The frequency of reported hydrological disasters has been increasing in recent years, as have their associated economic damages. The river system in Bangladesh is highly vulnerable to hydrological alterations in a changing climate, including:

(1) low flow situations becoming longer, more frequent, and more intense;
(2) flood peaks coming earlier, receding later, with more frequent small and large flood events; and
(3) increased flow variability.

In spite of recent declines in the human death toll and suffering from floods in Bangladesh, the economic damage has increased by several-fold, which underscores the role of economic development in amplifying infrastructural vulnerabilities. The changes in the hydrological regime—in the magnitude and seasonality characteristics of the flow system—may also trigger species extinction due to unfavourable spawning environments and impeded migration routes. Moreover, agricultural production losses may be caused by low flow situations, early flood incidences, and later recession of floods.

Under these circumstances, ecosystem-based adaptation options need to be explored extensively to reduce residual and net loss and damage. At the same time, the trade-off between securing agricultural production and ecosystem conservation needs to be balanced effectively. An integrated basin-wide approach is long due to protect the health of the transboundary rivers in the greater GBM Basins region. Inaction and delayed action may continue to cause irreversible loss and damage of ecosystem components and services.

Box 3.1: The Water-Food-Climate Nexus – implications for long term adaptive planning such as the Bangladesh Delta Plan by Catharien Terwisscha van Scheltinga.[1]

Water management of the future starts with planning today. This is a highly uncertain and very complex matter. The Bangladesh Delta Plan 2100 aims to assist in this longer-term planning, and to facilitate long term economic development through improving food security, water safety and availability. An adaptive plan is developed and side by side implementation is started to improve future land and water management, in relation to water safety and food security.

The Bangladesh Delta Plan aims at adaptive planning. It focuses on bringing various strategies in the picture, and optimizing the interventions and investments taking into account climate change and other uncertainties (Box Fig. 3.5).

---

[1]Bangladesh Delta Plan Technical Assistance Team. Contributions of BDP colleagues Dewan Abdul Quadir, Zahurul Karim, Giasuddin Choudhury, Zahirul Haque Khan, Professor Shamsul Alam, Taibur Rahman, Mofidul Islam, Mirzanur Rahman, Fulco Ludwig, Maaike van Aalst, William Oliemans and Jaap de Heer are kindly acknowledged.

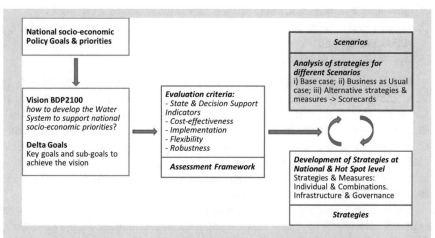

Fig. 3.5 Schematic of the Bangladesh Delta Plan. *Source* Catharien Terwisscha van Scheltinga

Climate change has large impacts on both water and agricultural systems in Bangladesh. Climate change affects water available for irrigation but also directly influences plant growth. Land use changes to improve agricultural production have a large impact on water management both in terms of flood protection and to supply water for food production. In addition, in the southwest, climate change and upstream developments have a large impact on salt water intrusion. This salt intrusion causes large scale changes in the agricultural production systems. In the development of the delta plan this close relationship between climate change adaptation, food security and water management is taken into account, using a hotspot approach: in each of the hotspots, a strategy is being formulated, based on which measures for future development can be formulated. In the course of 2016 it is expected that further information will be available regarding the strategies. Infrastructure, knowledge development and institutional changes will go side by side in order to ensure food security for the country.

So far, national goals have been translated into a vision BDP2100 with delta goals. Together with scenarios for delta management which have been developed,[2] and the strategies for the different hotspots currently being developed, this provides a basis to come to an assessment of strategies and measures for the future.

---

[2]Maaike van Aalst, William Oliemans, Fulco Ludwig, Catharien Terwisscha Scheltinga and Kymo Slager, with contributions from the BDP2100 team (2015), *Process of scenario development and draft scenarios for the BDP2100.*

References

Agrawala S, Ota T, Ahmed AU, Smith J, van Aalst M. 2003. *Development and Climate Change in Bangladesh: Focus on Coastal Flooding and the Sundarbans*. COM/ENV/EPOC/DCD/DAC (2003)3/FINAL. OECD.

Ahmed AU. 2005. Adaptation options for managing water-related extreme events under climate change regime: Bangladesh perspectives. In: Mirza MMQ, Ahmad QK (eds). *Climate Change and Water Resources in South Asia*. London, UK: A.A. Balkema Publishers. 255–278.

Bates B, Kundzewicz Z, Wu S, Palutikof J (eds). 2008. *Climate Change and Water*. Technical Paper of the Intergovernmental Panel on Climate Change. Geneva: IPCC Secretariat.

Bangladesh Bureau of Statistics (BBS). 2016. *National Accounts Statistics*. Dhaka: Bangladesh Bureau of Statistics, Statistics and Informatics Division, Ministry of Planning.

Brammer H. 1983. Agriculture and food production in polder areas. *Water International* 8: 74–81.

Brouwer R, Akter S, Brander L, Haque E. 2007. Socioeconomic vulnerability and adaptation to environmental risk: a case study of climate change and flooding in Bangladesh. *Risk Analysis* 27: 313–326.

Center for Environmental Geographic Information Systems (CEGIS). 2012. *Master plan of haor areas*. Dhaka: Bangladesh Haor and Wetland Development Board, Ministry of Water Resources, Government of the Peoples Republic of Bangladesh.

Chow J. 2017. Mangrove management for climate change adaptation and sustainable development of coastal zones. *Journal of Sustainable Forestry*. https://doi.org/10.1080/10549811.2017.1339615.

Faruque HSM, Ali ML. 2005. Climate change and water resources management in Bangladesh. In: Mirza MMQ, Ahmad QK (eds). *Climate Change and Water Resources in South Asia*. London, UK: A.A. Balkema Publishers. 231–254.

Gain AK, Apel H, Renaud FG, Giupponi C. 2013. Thresholds of hydrologic flow regime of a river and investigation of climate change impact–the case of the Lower Brahmaputra river Basin. *Climatic Change* 120: 463–475.

Guha-Sapir D, Below R, Hoyois P. 2017. *EM-DAT*. www.emdat.be.

Hoque MA, Hoque MM, Ahmed KM. 2007. Declining groundwater level and aquifer dewatering in Dhaka metropolitan area, Bangladesh: causes and quantification. *Hydrogeology Journal* 15: 1523–1534.

Islam AS, Bala SK, Hussain MA, Hossain MA, Rahman MM. 2010. Performance of coastal structures during Cyclone Sidr. *Natural Hazards Review* 12: 111–116.

Jain SK. 2008. Impact of retreat of Gangotri glacier on the flow of Ganga River. *Current Science* 95: 1012–1014.

Jauhari VP, Gadhalay S. 2011. *Possible impact of climate change on India*. Hyderabad: Center For Climate Change and Environment Advisory.

Jiménez Cisneros BE, Oki T, Arnell NW, Benito G, Cogley JG, Döll P, Jiang T, Mwakalila SS. 2014. Freshwater resources. In: Field CB, Barros VR, Dokken DJ, Mach KJ, Mastrandrea MD, Bilir TE, Chatterjee M, Ebi KL, Estrada YO, Genova RC, Girma B, Kissel ES, Levy AN, MacCracken S, Mastrandrea PR, White LL (eds). *Climate Change 2014: Impacts, Adaptation, and Vulnerability. Part A: Global and Sectoral Aspects*. Contribution of Working Group II to the Fifth Assessment Report of the Intergovernmental Panel on Climate Change. Cambridge University Press, Cambridge, United Kingdom and New York, NY, USA, pp. 229–269.

Khalil GM. 1990. Floods in Bangladesh: a question of disciplining the rivers. *Natural Hazards* 3: 379–401.

Klein RJT, Midgley GF, Preston BL, Alam M, Berkhout FGH, Dow K, Shaw MR. 2014. Adaptation opportunities, constraints, and limits. In: Field CB, Barros VR, Dokken DJ, Mach KJ, Mastrandrea MD, Bilir TE, Chatterjee M, Ebi KL, Estrada YO, Genova RC, Girma B, Kissel ES, Levy AN, MacCracken S, Mastrandrea PR, White LL (eds). *Climate Change 2014: Impacts, Adaptation, and Vulnerability. Part A: Global and Sectoral Aspects*. Contribution of Working Group II to the Fifth Assessment Report of the Intergovernmental

Panel on Climate Change. Cambridge University Press, Cambridge, United Kingdom and New York, NY, USA, pp. 899–943.

Klein Tank AMG, Peterson TC, Quadir DA, Dorji S, Zou X, Tang H, Santhosh K, Joshi UR, Jaswal AK, Kolli RK, Sikder AB, Deshpande NR, Revadekar JV, Yeleuova K, Vandasheva S, Faleyeva M, Gomboluudev P, Budhathoki KP, Hussain A, Afzaal M, Chandrapala L, Anvar H, Amanmurad D, Asanova VS, Jones PD, New MG, Spektorman T. 2006. Changes in daily temperature and precipitation extremes in central and south Asia. *Journal of Geophysical Research* 111. D16105, https://doi.org/10.1029/2005jd006316.

Kundsewicz ZW, Kanae S, Seneviratne SI, Handmer J, Nicholls N, Peduzzi P, Mechler R, Bouwer LM, Arnell N, Mach K, Muir-Wood R, Brakenridge GR, Kron W, Benito G, Honda Y, Takahashi K, Sherstyukov B. 2014. Flood risk and climate change: global and regional perspectives. *Hydrological Sciences Journal* 59: 1–28.

Ministry of Environment and Forests (MoEF). 2005. *National Adaptation Plan of Action (NAPA)*. Dhaka, Bangladesh: Ministry of Environment and Forest, Government of the Peoples Republic of Bangladesh.

Ministry of Environment and Forests (MoEF). 2009. *Bangladesh Climate Change Strategy and Action Plan (BCCSAP)*. Dhaka, Bangladesh: Ministry of Environment and Forest Government of the Peoples Republic of Bangladesh.

Ministry of Water Resources (MoWR). 1999. *National Water Policy*. Dhaka, Bangladesh: Ministry of Water Resources, Government of the Peoples Republic of Bangladesh.

Mirza MMQ. 1998. Diversion of the Ganges water at Farakka and its effects on salinity in Bangladesh. *Environmental Management* 22: 711–722.

Mirza MMQ, Burton I. 2005. Using the adaptation policy framework to assess climate risks and response measures in South Asia: the case of floods and droughts in Bangladesh and India. In: Mirza MMQ, Ahmad QK (eds). *Climate Change and Water Resources in South Asia*. Taylor & Francis, London: 279–313.

Mirza MMQ, Ahmad QK. 2005. Climate change and water resoures in South Asia: an Introduction. In: Mirza MMQ, Ahmad QK (eds). *Climate Change and Water Resources in South Asia*. Taylor & Francis, London: 1–21.

Mirza MMQ, Warrick RA, Ericksen NJ, Kenny GJ. 2001. Are floods getting worse in the Ganges, Brahmaputra and Meghna basins? *Global Environmental Change Part B Environmental Hazards* 3: 37–48.

Mirza MMQ, Warrick RA, Ericksen NJ. 2003. The implications of climate change on floods of the Ganges, Brahmaputra and Meghna Rivers in Bangladesh. *Climatic Change* 57: 287–318.

Neumayer E, Barthel F. 2011. Normalizing economic loss from natural disasters: a global analysis. *Global Environmental Change* 21: 13–24.

Nishat A, Huq SMI, Barua SP, Reza AHMA, Khan ASM, Moniruzzaman AS (eds). 2002. *Bio-ecological zones of Bangladesh*. Dhaka, Bangladesh: International Union for Conservation of Nature and Natural Resources.

Nishat A, Khan MFA, Mukherjee N. 2011. *Assessment of Investment and Financial Flows to Adapt to the Climate Change Effects in the Water Sector UNDP Global Project*, Dhaka, Bangladesh: Government of the People's Republic of Bangladesh.

Nowreen S, Jalal MR, Khan MSA. 2014. Historical analysis of rationalizing South West coastal polders of Bangladesh. *Water Policy* 16: 264–279.

Paul BK, Rashid H, Islam MS Hunt LM. 2010. Cyclone evacuation in Bangladesh: tropical cyclones Gorky (1991) vs. Sidr (2007). *Environmental Hazards* 9: 89–101.

Rahman A, Dragoni D, El-Masri B. 2011. Response of the Sundarbans coastline to seal level rise and decreased sediment flow: a remote sensing assessment. *Remote Sensing of the Environment* 115: 3121–3128.

Safiuddin M, Karim MM. 2001. *Groundwater arsenic contamination in Bangladesh: causes, effects and remediation*. Dhaka, Bangladesh, In Proceedings of the 1st IEB International Conference and 7th Annual Paper Meet, The Institution of Engineers, Chittagong Center, Bangladesh.

Shahid S, Behrawan H. 2008. Drought risk assessment in the western part of Bangladesh. *Natural Hazards* 46: 391–413.

Turner AG, Annamalai H. 2012. Climate change and the South Asian summer monsoon. *Nature Climate Change* 2: 587–595.

Tsai C, Islam MN, Karim MR, Rahman KUMS. 1981. Spawning of major carps in the Lower Halda River, Bangladesh. *Estuaries* 4: 127–138.

United Nations Framework Convention on Climate Change (UNFCCC). 2012. *A Range of Approaches to Address Loss and Damage*. Bonn, Germany: United Nations Framework Convension on Climate Change.

Water Resources Planning Organization (WARPO). 2001. *National Water Management Plan*, Dhaka, Bangladesh: Water Resources Planning Organisation.

Chapter 4
Forest Management for Climate Change Adaptation in Bangladesh

Jeffrey Chow, Tanzinia Khanom, Riadadh Hossain
and Jennifer Khadim

Abstract Tropical forest management has a vital role to play in improving local adaptive capacity to climate change. Although forests are themselves often vulnerable to the consequences of climate change – such as the increased risk of fires or sea level rise – concerted management can also enhance the resilience of local communities to climate change-related events such as storm surges, coastal erosion, and landslides. This chapter explores the many ways that strategic forest management plays a role in climate change adaptation in Bangladesh, reviewing recent initiatives for conservation of the Sundarbans Reserve Forest, coastal afforestation, and reforestation of hill tract forests. Community-based social forestry, livelihood diversification, as well as disaster management are also key components of many of these programs. However, because baseline data and continual monitoring have been sparse, important uncertainties concerning long-term project effectiveness remain. Additional research is therefore necessary to evaluate the successfulness of forest-based adaptation in Bangladesh.

Keywords Forests · Management · Sunderbans · Afforestation
Disasters

Jeffrey Chow, International Centre for Climate Change and Development, Dhaka, Bangladesh, Corresponding Author, e-mail: jchow.conservation@gmail.com.

Tanzinia Khanom, International Centre for Climate Change and Development, Dhaka, Bangladesh.

Riadadh Hossain, International Centre for Climate Change and Development, Dhaka, Bangladesh.

Jennifer Khadim, International Centre for Climate Change and Development, Dhaka, Bangladesh.

© Springer Nature Switzerland AG 2019
S. Huq et al. (eds.), *Confronting Climate Change in Bangladesh*,
The Anthropocene: Politik—Economics—Society—Science 28,
https://doi.org/10.1007/978-3-030-05237-9_4

4.1 Introduction

Due to its location at the transition between tropical and temperate zones, Bangladesh has a rich biological heritage. Forests cover approximately 9.8% of Bangladesh's 14.8 million ha of land (BFD 2007), and consist of three major types (Hossain 2001): tropical moist deciduous forests, tropical wet evergreen and semi-evergreen forests, and coastal mangroves. Due to their location in particularly climate-sensitive coastal and hilly areas, coastal mangroves and tropical evergreen forests have an especially important role in adaptation. This chapter provides an overview of forest management as a climate change adaptation strategy, first by describing the role of forests in reducing climate-related damages, and then by discussing forest-related adaptation initiatives.

Mangroves, coastal forests that thrive at interfaces between land and sea in the tropics and subtropics, can provide services that would help reduce damage from climate change-induced sea level rise and resultant coastline instability (FAO 2007). Mangroves can rapidly colonize intertidal sediments and promote further sedimentation, as their extensive root structures help keep soils compact and slow coastal erosion (Blasco et al. 1996). Coastal plantations in Bangladesh have been used to accelerate land accretion and stabilize 120,000 hectares of coastland, with silt deposits of up to 3 m observed following 9–10 months of planting (Saenger/ Siddigi 1993; Iftekhar/Islam 2004).

Bangladesh has also been undertaking restoration and preservation of coastal greenbelts as protection against tropical cyclones, which are expected to become more intense due climate change and sea level rise (Karim/Mimura 2008). Mangroves' thickly grown leaves and dense networks of trunks, branches, and above-ground roots render them more resilient to storm surges than non-mangrove tree species, creating drag forces that significantly dissipate cyclonic storm wave energies (Mazda et al. 2006; Quartel et al. 2007). However, greenbelts can provide protection against intense storms only if appropriately designed and managed, while degraded mangroves can provide less shelter than expected (Mazda et al. 1997; Dahdouh-Guebas et al. 2005).

Although the vast majority of Bangladesh is low-lying floodplain, approximately 18% is hilly and prone to erosion and landslides (Islam/Uddin 2002; Mahmood/ Khan 2010). Landslide occurrence is related to geo-environmental characteristics such as terrain morphology, geology, land use, and vegetation. Increased landslide activity due to greater precipitation is an expected impact from climate change (Crozier 2010). In Bangladesh, increased incidence of high-intensity rainfall due to climate change exacerbates landslide risk. The roots of tropical evergreen forests reduce erosion and enhance slope and topsoil stability, whereas deforested areas are more prone to landslides (Mahmood/Khan 2010).

Bangladesh has been undertaking conservation and afforestation of mangrove forests for decades, and in recent years the reforestation of hill forests has received attention as well. Livelihood diversification has been a major component of several of these projects, since communities in South Asia that are heavily dependent economically on both agriculture and forests can be especially vulnerable to climate change (Sushant 2013).

The following sections provide an overview of some major recent adaptation initiatives in Bangladesh in which forest management plays a key role. Section 4.2 describes projects related to conservation of the Sundarbans Reserve Forest in Bangladesh, part of the largest continuous expanse of mangroves in the world. Section 4.3 discusses plantation activities in the coastal regions, as well as in the hill forests. Section 4.4 concludes the chapter with a discussion of additional research needs.

4.2 Conservation of the Sundarbans

The Sundarbans is the largest contiguous mangrove ecosystem in the world and produces a diverse range of ecosystem services (Mitchell 1995). A UNESCO World Heritage Site, the *Sundarbans Reserve Forest* (SRF) constitutes more than 50% of the forest area in Bangladesh and accounts for nearly 41% of the country's total forest revenue (Islam 2010; BFD 2010) (Fig. 4.1). The *Sundarbans Impact Zone* (SIZ) spans 5 districts in southwestern Bangladesh (O'Donnell and Wodon 2015). Approximately 3.5 million people directly or indirectly rely on the ecosystem services from the Sundarbans (Giri et al. 2007). Climate change is expected to cause sea level rise and salinity intrusion in the region, negatively impacting the ecosystem, exacerbating current conservation challenges, and threatening the livelihoods of surrounding communities (Agrawala et al. 2003; Loucks et al. 2010; Raha et al. 2012). Due to the importance of the Sundarbans in providing numerous ecosystem services, including those related to climate change adaptation, several projects have been attempted to conserve them.

The *Sundarbans Biodiversity Conservation Project* (SBCP) was a US$77.3 million project undertaken by *the Government of Bangladesh* (GoB), with the *Ministry of Environment and Forests* (MoEF) and the *Bangladesh Forest Department* (BFD) as executing agencies. The SBCP was mainly financed by a loan from the *Asian Development Bank* (ADB), grants from the *Global Environment Facility* (GEF), and the Dutch government. Commencing in 1999, the primary objective was to establish a sustainable management system for biodiversity conservation (ADB 2008). However, the ADB cancelled the project in 2005, two years prior to its scheduled completion date, due to delays, disagreements among stakeholders regarding project design, and other institutional issues. Post-project assessment has identified corruption and lack of local community engagement as additional reasons for failure (Hossain/Roy 2007). Despite its unsuccessful outcome, the SBCP demonstrated that implementation of large-scale conservation projects requires greater and earlier investment into policy and institutional reform, stakeholder buy-in, closer external supervision, as well as realistic, well-defined objectives (ADB 2008). These lessons have been relevant to more recent adaptation initiatives in Bangladesh.

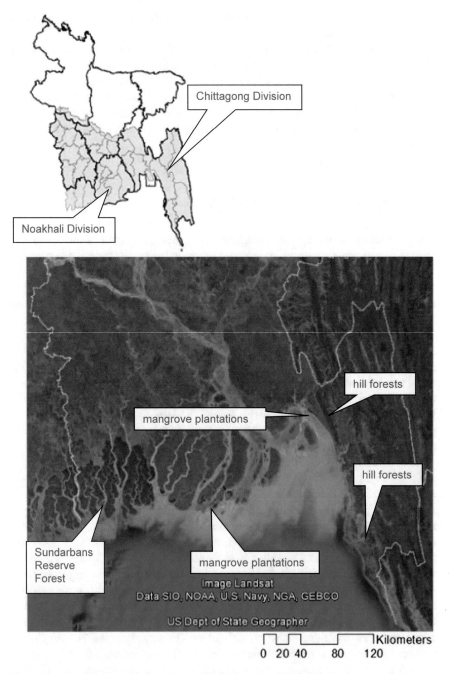

Fig. 4.1 Map of sundarbans, mangrove plantations, and hill forests in Bangladesh. Coastal districts in grey. *Source* Google Earth (2017); and authors own

The *Sundarbans Environmental and Livelihoods Security* (SEALS) project, funded by the European Union (EU) and implemented by the BFD, was a four-year program initiated in 2011. Given the failure of the SBCP, SEALS involved two key activities: (1) improving the capacity of GoB to protect and manage the SRF; and (2) reducing ecological stressors by promoting sustainable extraction of SRF resources and mitigating communities' dependence on them through developing alternative livelihoods (SEALS 2009).

One flagship project under the SEALS framework was the *Sundarbans Development and Alternative Resources Integration* (SUNDARI) project. Conducted by Concern Worldwide in collaboration with Shushilan, Jagrata Jubo Shanga, and the BFD, this project operated from 2012 to 2015 in the districts of Khulna and Satkhira. To reduce forest dependency and promote climate adaptive livelihoods, SUNDARI provided 15,000 *Sundarbans Dependent Extreme Poor Households* (SDEPHs) with skills training and seed money to help them adopt alternative income generating activities. Furthermore, the development of value chain mechanisms for SRF products improved access to markets. Through awareness and motivational campaigns, SUNDARI also helped promote communities' access to the government's Social Safety Net Programme. With livelihood diversification, about 14% of the SDEPHs had stopped collecting resources from the forest by 2014 (Sarwar 2015).

Additionally, as part of their climate change adaptation plan, SUNDARI has undertaken several disaster preparedness initiatives, including the establishment of mock cyclone drills and educational institutions. They have developed disaster management plans at the individual household level for over 701 SDEPHs and have promoted community engagement with local governments and disaster management committees. SUNDARI also hired 500 local workers, mostly women, to build an embankment that reduced flooding for 5000 people during the monsoon season. SUNDARI has developed a local model for integrated social forestry and aquaculture, based upon a grassroots co-management system, which has helped Sundarbans communities move towards sustainable forest management and biodiversity protection (Sarwar 2015).

In 2013, the *United States Agency for International Development* (USAID) launched the five-year, US$8 million Climate-Resilient Ecosystems and Livelihoods program, with the overarching aim of protecting the wetlands and forests of Bangladesh through co-management of natural resources by government ministries, technical agencies, and various community-based organizations (USAID 2013). The key activities under the program include promoting alternative, climate-resilient livelihood opportunities, guiding policies and processes for co-management of natural resources, and implementing the 2009 *Bangladesh Climate Change Strategy and Action Plan* (BCCSAP) (US Department of State 2014).

The EU funded another SRF conservation project called *Community-based Action Towards the Sustainability of the Sundarbans Reserve Forest* (CBAS-SRF). World Vision Bangladesh was the lead implementer for the project, which ran from 2012 to 2015 in the Bagerhat and Pirojpur districts. CBAS-SRF worked to improve the biodiversity of the forest, reduce the risk of disasters among SDEPHs, and develop alternative livelihoods for the targeted communities. As a community-based adaptation initiative, 565 men and 203 women were trained in disaster management. To reduce fuelwood extraction from the forest, 333 energy saving stoves were distributed among the beneficiaries as part of the project (Ghosh 2013).

These recent projects broadly follow the principles expressed in the Bangladesh Sundarban Delta Vision 2050, a vision document developed by *the International Union for Conservation of Nature* (IUCN) Bangladesh, and funded by the *World Wildlife Fund for Nature* (WWF) Netherlands and the Dutch embassy, to guide ecological and livelihood conservation in the SIZ (IUCN Bangladesh 2014). After consultation with a range of stakeholders, six key strategies have been highlighted for achievement by the year 2050: (1) improving the current governance structure; (2) establishing suitable mechanisms and information systems for sustainability of ecosystem services; (3) increasing freshwater flows into the delta; (4) reducing pollution; (5) developing alternative livelihoods for local communities; and (6) implementing measures to alleviate the negative impacts of climate change (IUCN Bangladesh 2014). Thus, recent conservation projects in the Sundarbans have largely focused on capacity building, improving the sustainability of resource extraction, developing alternative livelihoods, and disaster management.

4.3 Coastal and Hill Forest Plantations

Since 1966, the GoB has undertaken plantings of mangroves on newly accreted, or *char*, lands in the coastal districts within the Barisal and Chittagong Divisions in order to promote shoreline stabilization, land accretion, and storm surge protection (Iftekhar and Islam 2004) (Fig. 4.1). Two species, *Sonneratia apetala* and *Avicennia officinalis*, demonstrated promising survival rates and were planted in virtually monocultural stands, though some plantations included additional species. *S. apetala* was the preferred species, comprising over 80% of the plantations. In 1976, the Government of Bangladesh granted jurisdiction over 497,976 ha of *chars* to the BFD for mangrove afforestation (Islam 2000).

Since 1980, financing for afforestation has come largely from multilateral development institutions and other sources of international assistance, with the GoB providing support mostly in the form of in-kind contributions. From 1980 to 1992, the World Bank-funded Mangrove Afforestation Projects I and II committed over US$22 million to plant approximately 8100 ha of mangroves annually (Saenger/ Siddiqi 1993). The subsequent US$53 million Forest Resources Management Project, implemented between 1992 and 2001, included 33,570 ha of new mangrove plantations and 36,500 of industrial plantations (World Bank 2013).

By 2001, approximately 148,500 ha of mangroves, having survived initial pest infestations, were successfully established. However, erosion and conversion for agriculture and settlements following land stabilization have destroyed most plantations in the Chittagong Division, whereas the plantations in the Barisal Division are largely intact (Iftekhar/Islam 2004). In Noakhali especially, plantations resulted in rapid land accretion, creating areas suitable for farming. By 2007, about 45,000 ha of mature mangrove plantations remained (BFD 2007). 60 km of these forest belts are situated on the seaward side of coastal embankments, but this leaves over 90% of these embankments exposed (World Bank 2010).

Under current law, the Land Ministry of Bangladesh allocates authority over *chars* to the BFD for a period of 20–25 years for mangrove plantation (Iftekhar 2006; GoB/UNDP 2008). This policy requires that 25 years after establishment, half of the plantation lands remain in reserve forest status to provide ecosystem services, while the other half is returned to the Land Ministry, where it is then often redistributed to communities via local administrations (World Bank 2013). Forest clearing, often for farming or settlements, requires MoEF approval, though the rate of compliance is unknown.

The BFD has continued to undertake new coastal plantation activities and has recently started reforestation programs in the Chittagong hill districts, with both goals outlined in Bangladesh's Sixth Five Year Plan (GoB 2012). According to the BCCSAP and the expected National Adaptation Plan, afforestation and reforestation are also part of Bangladesh's efforts to mitigate climate change via carbon sequestration (Bangladesh MoEF 2009). Bangladesh will require an additional US $75 million devoted to forestry in order to adapt to climate change by 2050 (World Bank 2010). Working towards these objectives, the GoB has funded the reforestation of 4971 ha of barren land through its Bangladesh Climate Change Trust Fund (Kamruzzaman 2015). The afforestation of 300 ha on the seaward side of coastal embankments by 2020 is also an objective of the Coastal Embankment Improvement Project, funded via the Pilot Program for Climate Resilience as specified in Bangladesh's Strategic Program for Climate Resilience (GoB 2010; Forni 2015). Additionally, the GoB has embarked on two major projects: Community Based Adaptation to Climate Change through Coastal Afforestation in Bangladesh (CBACC-CAB) and the Climate Resilient Participatory Afforestation and Reforestation Project (CRPARP).

4.3.1 Community Based Adaptation to Climate Change Through Coastal Afforestation

The CBACC-CAB is a collaborative effort between the Government of Bangladesh and the United Nations Development Program to increase the resilience of coastal communities to climate change while also providing new options for livelihood generation (UNDP 2011). With its initial phase taking place from 2009 to 2014 and financed with US$3.3 million from the Least Developed Countries Fund administered by the GEF, CBACC-CAB contributes to creating a forest shelterbelt in the Meghna estuary as part of Bangladesh's National Adaptation Programme of Action for climate change (GoB 2007). In addition to mangrove afforestation, project objectives have included the establishment of roadside strips and dyke plantations of terrestrial tree species, and mound plantations to facilitate salinity-resistant agriculture. The project has also promoted an innovative ditch/dyke system of simultaneous agriculture, horticulture, and aquaculture, for the dual purposes of storm surge mitigation and livelihood diversification in coastal communities (Ferguson/Das 2012). The project also introduces new "model" multi-species plantations in order to encourage ecological succession.

Besides the BFD, other collaborators of the program include the *Bangladesh Forest Research Institute* (BFRI), Department of Agricultural Extension, Department of Fisheries, Department of Livestock Services, and the Bangladesh Water Development Board. Between 2009 and 2014, the BFD established 9000 ha of new mangrove plantations, as well as 332 ha of mound plantations, 112 ha of dyke plantations, and 680 km of strip plantations at four project sites in the coastal districts of Chittagong, Noakhali, Bhola, and Patuakhali (Paramesh Nandy, personal communication, 11 November 2015). The BFRI established 200 ha of model plantations. The other agencies have undertaken demonstration projects and trainings for various related agriculture, aquaculture, horticulture, forestry, and embankment projects that are connected to livelihood diversification and adaptation to increased sea levels, salinity, and storm events due to climate change. The project has reached over 30,000 households in coastal Bangladesh via these activities.

4.3.2 Climate Resilient Participatory Afforestation and Reforestation Project

In 2013 the World Bank approved US$35 million from the Bangladesh Climate Change Resilience Fund for the CRPARP, which focuses on afforestation and reforestation in both coastal and hilly districts (World Bank 2013). Implemented by the BFD and the Arannayk Foundation, the project devotes US$19 million directly to plantation activities in the coastal districts of Cox's Bazar, Chittagong, Noakhali, Laxmipur, Feni, Barisal, Patuakhali, Barguna, and Bhola, as well as, in the hilly areas, the core and buffer zones of the Reserved Forest Land of Chittagong and Cox's Bazar (Fig. 4.1). In addition to 5700 ha of new mangrove plantations, the project aims to plant 345 ha of mound plantations, 82 ha of dyke plantations, 410 ha of jhaw plantations, 260 ha of enrichment plantations, 635 km of golpata plantations, and 1037 km of road/embankment-side plantations (Uttam Kumar Saha, personal communication, 14 March 2015). In the hilly areas, the project focuses on undulating terrains between 15 and 20 km of the coastal zone, since these areas are also impacted by cyclones (World Bank 2013). The project intends to establish 3102 ha of plantations within the core zone and 6733 ha within the buffer zone, with another 400 ha of additional buffer plantation. These activities are expected to be complete by the end of 2016.

The project also includes a number of capacity building and education initiatives meant to foster local participation. The BFD engenders local engagement through hiring local villagers as laborers and training them in nursery and plantation techniques. The Arannayk Foundation, in turn, promotes alternative livelihoods to reduce collection of forest goods. Alternative income generating activities include, at the household level, livestock rearing, vegetable cultivation, and small-scale trade. Community level activities include establishing cooperatives to improve marketing of local products, eco-tourism, and making energy efficient cook stoves. This project component aims to reach 6000 households in 200 forest communities across the 9 project districts. Similar previous initiatives by the Arannayk Foundation have been found to reduce logging by 70%, reduce fuelwood collection by 50–60%, reduce bamboo and cane collection by 50–70%, and reduce swidden agriculture by 90% (World Bank 2013), thus reducing ecological stress on the forest and maintaining it for climate adaptive ecosystem services.

4.4 Needs Moving Forward

The preservation and plantation of mangroves and hill forests represent promising strategies to help communities in Bangladesh withstand the damaging impacts of climate change. When local ecosystem benefits are accounted for, the net value of forest management to communities in Bangladesh can exceed its costs, particularly when the costs of labour and other inputs are low (Chow 2015).

However, to date, mangrove promotion has been an imperfect adaptation strategy. Rehabilitation and restoration projects have had mixed results due to inadequate site selection, improper soil preparation and planting techniques, and insufficient biodiversity (Alongi 2002), which impede both establishment and regeneration. Furthermore, mangroves themselves are sensitive to potential changes in salinity and sea level rise (McLeod/Salm 2006). Structurally weak mangroves could harm local communities by providing a false sense of security against tropical storm surges. Adequate training of coastal managers, adaptable planting and replanting schemes, and rigorous monitoring and evaluation are therefore necessary for mangrove projects to successfully provide protection against climate damages. A USAID-funded project called Strengthening the Environment, Forestry and Climate Change Capacities of the Ministry of Environment and Forests and its Agencies aims to enhance human and organizational capacity to address these some of these issues (Budhathoki 2015).

Many important knowledge gaps remain that require continued investigation. For example, the necessary plantation to supply more than trivial storm protection, or landslide control, is largely unknown, with countries setting mandatory plantation widths somewhat arbitrarily (Iftekhar/Islam 2004). Moreover, the lack of any systematic and routine inventory and monitoring of forest resources in Bangladesh makes baseline data unavailable for project assessment (World Bank 2013). The forest inventory data which does exist is often sparse, unreliable, and subject to error. Consequently, despite the enthusiasm for forestry projects in Bangladesh, there is little empirical evidence to date that these conservation projects and plantations can be self-sustaining. The effectiveness of forest preservation and restoration in providing climate change adaptation benefits is thus a vital area for continual research.

References

Agrawala S, Ota T, Ahmed AU, Smith J, van Aalst M. 2003. *Development and Climate Change in Bangladesh: Focus on Coastal Flooding and the Sundarbans.* COM/ENV/EPOC/DCD/DAC (2003)3/FINAL. OECD.

Alongi DM. 2002. Present state and future of the world's mangrove forests. *Environmental Conservation* 29: 331–349.

Asian Development Bank (ADB). 2008. *ADB Completion Report: Bangladesh: Sundarbans Biodiversity Conservation Projec*t. Project Number: 30032. Loan Number: 1643. Asian Development Bank.

Bangladesh Forest Department (BFD). 2007. *National Forest and Tree Resources Assessment 2005–2007.* Bangladesh Ministry of Environment and Forest (MoEF) and Food and Agriculture Organization of the United Nations (FAO), Dhaka, Bangladesh.

Bangladesh Forest Department (BFD). 2010. *Integrated Resources Management Plans for The Sundarbans (2010–2020).* Volume 1. Nishorgo Network, Dhaka, Bangladesh.

Bangladesh Forest Department (BFD), Ministry of Environment and Forest, Bangladesh Space Research and Remote Sensing Organization, Ministry of Defense. 2007. *National Forest and Tree Resources Assessment 2005–200*7. Food and Agriculture Organization of the United Nations, Dhaka, Bangladesh.

Bangladesh Ministry of Environment and Forest (MoEF) Government of the People's Republic of Bangladesh. 2009. *Bangladesh Climate Change Strategy and Action Plan 2009. Ministry of Environment and Forests*, Government of the People's Republic of Bangladesh, Dhaka, Bangladesh.

Blasco F, Saenger P, Janodet E. 1996. Mangroves as indicators of coastal change. *Catena* 27: 167–178.

Budhathoki P. 2015. *Strengthening the Environment, Forestry and Climate Change Capacities of the Ministry of Environment and Forests and its Agencies (GCP/BDG/053/USA): Introduction to the Project, its Current Status and Future Plans.* Presentation at the First Meeting of the Technical Advisory Group, 5 February 2015. Dhaka, Bangladesh.

Chow J. 2015. Spatially explicit evaluation of local extractive benefits from mangrove plantations in Bangladesh. *Journal of Sustainable Forestry* 34: 651–681.

Crozier MJ. 2010. Deciphering the effect of climate change on landslide activity: a review. *Geomorphology* 124: 260–267.

Dahdouh-Guebas F, Jayatissa LP, Di Nitto D, Bosire JO, Lo Seen D, Koedam N. 2005. How effective were mangroves as a defence against the recent tsunami? *Current Biology* 15: R443–R447.

Ferguson A, Das R. 2012. *Mid-term Evaluation of Community Based Adaptation to Climate Change through Coastal Afforestation in Bangladesh.* Final Report. UNDP Bangladesh.

Food and Agriculture Organization of the United Nations (FAO). 2007. *The World's Mangroves 1980–2005.* FAO Forestry Paper 153. FAO, Rome, Italy.

Forni MS. 2015. *Bangladesh – Coastal Embankment Improvement Project – Phase I (CEIP-I): P128276 – Implementation Status Results Report: Sequence 05.* Washington, DC: World Bank Group. http://documents.worldbank.org/curated/en/2015/12/25717184/bangladesh-coastal-embankment-improvement-project-phase-ceip-i-p128276-implementation-status-results-report-sequence-05.

Ghosh A. 2013. *CBAS Project Annual Report 2012*. World Vision, Dhaka, Bangladesh.

Giri C, Pengra B, Zhu Z, Singh A, Tiezsen LL. 2007. Monitoring mangrove forest dynamics of the Sundarbans in Bangladesh and India using multi-temporal satellite data from 1973 to 2000. *Estuarine, Coastal and Shelf Science* 73: 91–100.

Government of Bangladesh (GoB). 2007. *National Adaptation Programme of Action*. People's Republic of Bangladesh, Ministry of Environment and Forest Government, Dhaka, Bangladesh.

Government of Bangladesh (GoB). 2010. *Bangladesh: Strategic Program for Climate Resilience (SPCR)*. Economic Relations Division, Ministry of Finance, Government of the People's Republic of Bangladesh, Dhaka, Bangladesh.

Government of Bangladesh (GoB). 2012. *Sixth Five Year Plan: FY2011–FY2015*. General Economics Division, Planning Commission, Ministry of Planning, Government of the People's Republic of Bangladesh, Dhaka, Bangladesh.

Government of Bangladesh (GoB) and the United National Development Programme (UNDP). 2008. *Community-based Adaptation to Climate Change through Coastal Afforestation in Bangladesh*. Project Document. PIMS 3875.

Hossain J, Roy K. 2007. *Deserting the Sundarbans: Local People's Perspective on ADB-GEF-Netherlands funded Sundarban Biodiversity Conservation Project*. Unnayan Onneshan–the Innovators, Dhaka, Bangladesh.

Hossain MK. 2001. Overview of the forest biodiversity in Bangladesh. In: *Assessment, Conservation and Sustainable Use of Forest Biodiversity*, CBD Technical Series no. 3. Secretariat of the Convention on Biological Diversity (ed.). SCBD, Montreal. 33–35.

Iftekhar MS. 2006. Conservation and management of the Bangladesh coastal ecosystem: overview of an integrated approach. *Natural Resources Forum* 30: 230–237.

Iftekhar MS, Islam MR. 2004. Managing mangroves in Bangladesh: a strategy analysis. *Journal of Coastal Conservation* 10: 139–146.

Islam KMN. 2010. *A Study of the Principal Marketed Value Chains derived from the Sundarbans Reserved Forest*. Integrated Protected Area Co-Management (IPAC) Volume 1: Main Report. USAID Bangladesh. Dhaka, Bangladesh.

Islam MT. 2000. *Integrated Forest Management Plan for the Chittagong Coastal Afforestation Division*. Government of Bangladesh/World Bank Forest Resources Management Project Technical Assistance Component. Khulna, Bangladesh.

Islam MN, Uddin MN. 2002. *Country Paper on Hydrogeology Section in International Workshop on Arsenic Issue in Bangladesh, 14–16 January, 2002*, http://phys4.harvard.edu/~wilson/arsenic/remediation/Reports/Countrypaper.doc.

IUCN Bangladesh. 2014. *Bangladesh Sundarban Delta Vision 2050 – Document 1: The Vision*. IUCN Bangladesh, Dhaka.

Kamruzzaman M. 2015. *National Climate Finance: Performance of Bangladesh Climate Change Trust Fund*. Presentation at Gobeshona Conference, 10 January 2015, Dhaka, Bangladesh.

Karim FK, Mimura N. 2008. Impacts of climate change and sea-level rise on cyclonic storm surge floods in Bangladesh. *Global Environmental Change* 18: 490–500.

Loucks C, Barber-Mayer S, Hossain MAA, Barlow A, Chowdhury RM. 2010. Sea level rise and tigers: predicted impacts to Bangladesh's Sundarbans mangroves. *Climatic Change* 98: 291–298.

Mahmood AB, Khan MH. 2010. *Landslide vulnerability of Bangladesh hills and sustainable management options: a case study of 2007 landslide in Chittagong City*. SAARC Workshop on Landslide Risk Management in South Asia. Thimpu, Bhutan. 61–71.

Mazda Y, Magi M, Ikeda Y, Kurokawa T, Asano T. 2006. Wave reduction in a mangrove forest dominated by *Sonneratia* sp. *Wetlands Ecology and Management* 14: 365–378.

Mazda Y, Magi M, Kogo M, Hong PN. 1997. Mangroves as coastal protection from waves in the Tong King delta, Vietnam. *Mangroves and Salt Marshes* 1: 127–135.

McLeod E, Salm RV. 2006. *Managing Mangroves for Resistance to Climate Change*. IUCN Resilience Science Group Working Paper Series No. 2. IUCN, Gland, Switzerland.

Mitchell A. 1995. *Integrated Resource Development of the Sundarbans Reserved Forest, Bangladesh: Draft Report on Natural Resource Economics*. FO:DP/BGD/84/056. United National Development Programme, Food and Agriculture Organization, Khulna, Bangladesh.

O'Donnell A, Wodon Q (eds). 2015. *Climate Change Adaptation and Social Resilience in the Sundarbans*. Routledge, New York.

Quartel S, Kroon A, Augustinus PGEF, Van Santen P, Tri NH. 2007. Wave attenuation in coastal mangroves in the Red River Delta, Vietnam. *Journal of Asian Earth Sciences* 29: 576–584.

Raha A, Das A, Banerjee K, Mitra A. 2012. Climate change impacts on Indian Sunderbans: a time series analysis (1924–2008). *Biodiversity and Conservation* 21: 1289–1307.

Saenger P, Siddiqi NA. 1993. Land from the sea: the mangrove afforestation program of Bangladesh. *Ocean and Coastal Management* 20: 23–29.

Sarwar GM. 2015. SUNDARI: *Protecting the biodiversity of the Sundarbans by reducing human pressure – Final Booklet*. Concern Worldwide, Bangladesh.

Sundarbans Environmental and Livelihoods Security (SEALS). 2009. *Formulation Study for the Sundarbans Environmental and Livelihoods Security (SEALS) Project*. AGRIFOR Consult, Dhaka, Bangladesh.

Sushant. 2013. Impact of climate change in eastern Madhya Pradesh, India. *Tropical Conservation Science* 6: 338–364.

United Nations Development Program (UNDP). 2011. *Bangladesh Case Study: Community Based Adaptation to Climate Change through Coastal Afforestation in Bangladesh (CBACC-CF Project)*. United Nations Development Program, Dhaka, Bangladesh.

US Department of State. 2014. *Meeting the Fast Start Commitment – U.S. Climate Finance in Fiscal Year 2012*.

USAID. 2013. USAID launches new environment initiative to improve Bangladesh's resilience to Climate Change [Press Release].

World Bank. 2010. *The Economics of Adaptation to Climate Change: A Synthesis Report*. Washington DC: World Bank, Washington, DC.

World Bank. 2013. *Bangladesh – Climate Resilient Participatory Afforestation and Reforestation Project*. World Bank, Washington, DC.

Chapter 5
Ecosystem-Based Adaptation: Opportunities and Challenges in Coastal Bangladesh

M. Mustafa Saroar, M. Mahbubur Rahman, Khalid M. Bahauddin and M. Abdur Rahaman

Abstract *Ecosystem-based adaptation* (EbA) is the use of biodiversity and ecosystem services as part of a climate change adaptation strategy. In coastal Bangladesh, EbA is implemented to enhance the resilience of coastal agriculture, fisheries, forestry, and settlements against the impacts of both climatic and non-climatic stressors. This chapter discusses the current status and challenges of EbA adoption in coastal Bangladesh. First it presents a succinct review of the climate change impacts which motivate EbA adoption. It then describes major types of EbA, which are linked to cropping practices, soil and nutrient management, water management, erosion control, and food and livelihood security. Finally, it proposes integrated institutional approaches to bolster EbA's potential. Existing EbA strategies should conform to scientific knowledge, which would help improve community resilience and ecosystem health.

Keywords Ecosystems · Biodiversity · Agriculture · Crops · Livelihoods

5.1 Introduction

Coastal Bangladesh is comprised of 19 districts in three regions. The south-east region covers Chittagong, Cox's Bazar, and St. Martin's Island. The south-central region includes Chandpur, Bhola, Patuakhali, Noakhali, and Barguna. The south-west region encompasses Khulna, Bagerhat, and Satkhira. Mostly situated one meter above mean sea level, the coastal ecosystems of Bangladesh are very

M. Mustafa Saroar, Department of Urban & Regional Planning, Khulna University of Engineering & Technology (KUET), Khulna 9203, Bangladesh, Corresponding Author, e-mail: saroar.mustafa@yahoo.com.

M. Mahbubur Rahman, Network on Climate Change, Bangladesh (NCC,B) Trust, Dhaka, Bangladesh.

Khalid M. Bahauddin, International Society for Sustainability.

M. Abdur Rahaman, Climate Change Adaptation, Mitigation Experiment & Training (CAMET) Park, Noakhali, Bangladesh.

© Springer Nature Switzerland AG 2019
S. Huq et al. (eds.), *Confronting Climate Change in Bangladesh*,
The Anthropocene: Politik—Economics—Society—Science 28,
https://doi.org/10.1007/978-3-030-05237-9_5

diverse, providing significant services to more than 35 million people. However, the entire coastal zone is subject to several types of hydro-meteorological disasters, such as floods, cyclones, tidal surges, and salinity intrusion. Coastal ecosystems also face several non-climatic stressors, including land use change, sedimentation, soil erosion, water pollution, and overfishing (Saroar et al. 2015). For example, over the last three decades the unplanned expansion of saltwater shrimp aquaculture – often by clearing rice fields, wetlands, and mangroves – has resulted in habitat destruction and salinization of around 1.5 million hectares of land (Paul/Vogl 2011). The impacts of these climatic and non-climatic stressors, particularly on biodiversity, are complex and often site-specific (Dickinson et al. 2015).

Following the IPCC (2007) Fourth Assessment Report, the *Institute for Water Modeling* (IWM) and the *Center for Environmental Geographic Information Systems* (CEGIS) (2007) projected that approximately 13 percent more land (469,000 ha) in the coastal zone of Bangladesh will experience regular flooding if sea levels rise 62 cm by 2080, according to the A2 emissions scenario under which the world is around 3 °C warmer by 2100. An additional 16 percent (551,500 ha) will be waterlogged if rainfall synchronizes with this *sea level rise* (SLR). In addition to coastal flooding and waterlogging, under a 2 °C warming scenario, SLR and greater tropical storm wind speeds may increase tidal surge heights by up to 21% (Ali 1999).

Traditional adaptation strategies, predominantly infrastructural measures such as shelters and dykes, have lowered the human death toll from major disasters, but they do not help protect coastal ecosystems and biodiversity. *Ecosystem-based adaptation* (EbA), which is the use ecosystem services as part of an overall adaptation strategy (MEA 2005), can help people adapt to the adverse effects of climatic and non-climatic stressors (Munang et al. 2013). EbA can enhance coastal resilience through numerous pathways including protection against cyclonic storm surges; erosion control and soil management; pollution control and water purification; irrigation and drainage improvement; biodiversity conservation; and food security. The following section focuses on their role in coastal ecosystem protection and livelihood security. The effectiveness of EbA for building coastal resilience against climatic and non-climatic stressors is constrained by numerous institutional and policy challenges (Doswald/Estrella 2015), which are explored in the remainder of this chapter.

5.2   Status and Role of EbA in Coastal Bangladesh

Following the publication of the IPCC's (2001) Third Assessment Report, which describes Bangladesh as one of the countries most vulnerable to climate change, Bangladesh drafted its *National Adaptation Programme of Action* (NAPA) (GoB 2005a) and Bangladesh Climate Change Strategy and Action Plan (GoB 2008), which paved the way for increased involvement of government and *non-governmental organizations* (NGOs) in EbA. This section discusses EbA strategies which may be grouped into two broad categories: coastal ecosystem protection and the development of climate adaptive agriculture.

5.2.1   Coastal Ecosystem Protection

The coastal zone of Bangladesh encompasses several ecosystem types, including mangroves, wetlands, natural canals, and floodplains, which are vulnerable to both climatic and non-climatic stressors (Nandy et al. 2013). Due to the site specificity of these risks, successfully addressing these stressors through EbA requires the careful assessment of exposure and sensitivity of the ecosystem's components to these stressors. Furthermore, because coastal ecosystems interact across many sectors, EbA needs an integrated approach.

Unfortunately, some sectoral management plans to protect and manage coastal ecosystems have failed to address these issues. For instance, the National Water Management Plan (GoB 2001) emphasizes river basin management and coastal embankments, but gives very cursory attention to wetlands conservation (Iftekhar 2006). Embankment programs have been relatively successful in mitigating salt-water intrusion, but in some cases polders, the embanked areas, suffer from excessive waterlogging (Saroar et al. 2015). However, with the adoption of Bangladesh's Coastal Zone Policy of 2005 (GoB 2005b), the *integrated coastal zone management* (ICZM) approach – which promotes collaboration across sectors and agencies, participation and co-management by local stakeholders, and ecosystem-based and adaptive management (Schmidt/Duke 2015) – has been the Government of Bangladesh's primary strategy. This policy commits different ministries, departments and agencies to coordinate their activities in the coastal zone and elaborates the basis for EbA adoption across sectors.

Mangrove plantation as an EbA measure has been implemented in the south-central and south-west parts of coastal Bangladesh. Among other ecosystem services, mangroves can provide a line of defense for coastal communities, infrastructure, and livelihood assets against cyclonic storm surges, erosion, and salinity intrusion, particularly where embankments alone have been inadequate (Uddin et al. 2013). Although many mangrove plantations are large-scale projects funded by international donors and implemented by the Government of Bangladesh, there are small-scale community-based plantations. For instance, the local NGO Caritas Bangladesh and the Bangladesh Forest Department have implemented, in Shyamnagar Upazila, Satkhira District, a small scale green belt program, which helps to protect embankments and reduce soil erosion (Iftekhar 2006; Iftekhar/Takama 2008). Moreover, mangrove afforestation has facilitated long-term land use and ownership through the stabilization of newly accreted lands (Nandy et al. 2013). For a more thorough examination of afforestation-centered EbA measures in Bangladesh, see Chap. 5 of this volume.

The coastal area also includes hundreds of natural canals and wetlands, which are used to drain excess water during the monsoon and retain water during the dry season. The canals and khas (government owned) wetlands are typically common pool resources used by communities for irrigating crops and for subsistence and commercial fishing. Increased monsoon rainfall causes drainage congestion and erosion, degrading the wetland and reducing their ecosystem service provision (Islam 2016). Following the EbA approach, local communities, with the help of NGOs and local administrations, in the south-west coast have been restoring these degraded canals to build community resilience against climate change. Restoration measures include re-excavation and bank stabilization by planting perennial plants and fruit trees, which controls erosion by maintaining vegetation cover on the dykes of the canals. For example, in JeleKhali village of Shyamnagar, a saline-prone area of Satkhira District, the local community, with assistance from Caritas Bangladesh, has restored a 5 km long degraded canal used for retaining rainwater for both irrigation and freshwater fisheries.

Other wetlands in Bangladesh include floodplains that foster one of the world's richest and most complex fisheries (see Box 5.1), as well as both cultivated and wild food plants (Sultana/Thompson 2008). The coastal floodplains are increasingly threatened by cyclonic surges, salt water intrusion and water pollution from urban and industrial sources (Hossain et al. 2015). One EbA measure is to create earthen embankments, stabilized by vegetation, that surround the wetlands. This helps restore wetland characteristics and build resilience to both climatic and non-climatic stressors by limiting soil erosion and preventing degradation caused by non-point source pollution, such as residual chemicals and fertilizers from agricultural operations (Rahman/Salehin 2013).

5.2.2 Adaptive Agriculture for Livelihood Security

Some parts of south-central and south-western coastal districts are affected by drainage congestion linked to both climatic and non-climatic factors (e.g. unplanned conversion of rice fields into aquaculture enclosures and dykes) (Islam/Wahab 2005; Islam 2016). Drainage congestion can sometimes create fishing opportunities, but it does not fully compensate for the loss of agricultural land (Mamun 2010). In the low-lying parts of the south-west coast, local NGOs, either in partnership with the *Bangladesh Center for Advanced Studies* (BCAS) or independently, have adopted floating hydroponic agriculture to create new opportunities for income, food, and livelihood security in waterlogged conditions. This EbA, a soilless agricultural practice, involves: (1) creating floating mats of water hyacinth and other aquatic vegetation available in wetlands during the early monsoon; and (2) laying down organic manure – usually cow dung and other materials – to make beds for growing vegetables. The practice replaces soil-based agriculture in waterlogged conditions, especially in wetland environments. Farmers are now not only able to produce short-rotation vegetable varieties to support livelihood and food security, but also reduce wetland degradation previously caused by chemical use, water pollution, and soil erosion (Nishat/Mukherjee 2013). With more land likely to be submerged due to increased rainfall, flooding, and SLR as a result of climate change, more people may be forced to adopt hydroponic agriculture as an EbA strategy.

In heavier monsoons, poor drainage negatively affects rain-fed amon rice, the main production crop. To adapt, the local people in Perojpur, Bagerhat and Barisal have excavated trenches alongside rice fields to hold excess rainwater, while raising the adjacent dykes to prevent rainwater from entering from neighboring lands. They cultivate rice inside the field, a variety of vegetables and fruit trees on the raised dykes, as well as fish in the trenches. This has not only contributed to farmers' nutritional security but also helped to prevent soil nutrient loss and erosion. Similar types of ecosystem-based integrated farming systems, called ditch-and-dyke schemes, have been implemented in Barguna, Noakhali, and Bhola (see Chap. 5). This EbA will be increasingly useful since monsoonal rainfall will intensify with climate change, causing drainage congestion in additional coastal sites.

Additionally, almost one million hectares of coastal agricultural land are affected by varying degrees of salinity. Depending on rainfall and salinity gradients, farmers have developed various portfolios of EbA approaches (PDO-ICZMP 2004). In the medium (8–12 ds/m) and high (>12 ds/m) saline areas, people have started growing amon rice during the monsoon season when the salinity level subsides. In the low to medium saline areas (0–8 ds/m) farmers have adopted BRRI Dhan-47 and BINA-8 rice during the dry season because of their resilience to salinity, moisture stress, and drier weather (Nandy et al. 2013). In flood free zones, farmers have also adopted non-rice crops such as maize, oil seeds, soybean, bean, bitter gourd, sweet gourd, chili, ground nut, and watermelon due to their resistance to high salinity during the dry season (CSISA 2012). A summary of additional EbA measures undertaken in coastal Bangladesh is presented in Table 5.1.

5.3 Challenges to the Adoption of EbA

EbA strategies have tremendous potential for building the resilience of communities, but are constrained by institutional factors and insufficient local level actions. Some EbA approaches are very narrowly focused, sometimes failing to integrate conservation and development. For example, some standalone initiatives, such as the Tidal River Management project, aim to trap sediment within polders, but do not allow fishing or cultivation (Islam/Kitazawa 2013; Haque et al. 2014). Although this helps to stabilize coastal land, it neglects the ecological and social processes that determine the livelihood security of economically marginalized populations (Mallick et al. 2011; Ahammad et al. 2013).

Successful implementation requires the engagement of various stakeholders, including local leaders and communities, elected representatives, and civil society organizations in designing and institutionalizing EbA (Vignola et al. 2009). However, many EbA projects have failed to ensure participation of relevant stakeholders. For example, the top-down selection of beneficiaries for coastal afforestation projects has ignored the specific needs of local communities. Similarly, although wetland and natural canal restoration projects have promoted soil stabilization and biodiversity conservation, such projects have also, in some cases, diminished livelihood opportunities from open water fishing. For instance, community-based fisheries management projects in Bangladesh have improved wetland restoration, conservation, and management. However, these projects have imposed fishing bans during the dry season and in deeper parts of wetland to protect breeding stock, without adequately establishing alternative livelihood opportunities (Sultana/ Thompson 2007). Also, ditch-and-dyke projects are seldom owned by the local community and thus their sustainability is difficult to ensure (Reid/Alam 2014).

Table 5.1 Summary of EbA measures undertaken in coastal Bangladesh

Impact area	EbA measures	Description	Ecosystem Functions and Services	Spatial coverage
Coastal ecosystem	Mangrove afforestation and conservation	Plantations of viable mangrove species are established in areas that are newly formed, accreted, or cleared in the past	Provides a natural buffer against coastal erosion, inundation, cyclonic storms Improves livelihoods of people dependent on the mangrove forest Provides coastal habitat for many plants, animals, and aquatic species	Coastal Barguna, Bhola, Noakhali and Perojpur districts
	Revegetation in the newly formed land	Degraded coastal land is kept fallow for years to recover, usually in areas severely affected by cyclone and tidal surges with return periods of several years	Re-establishes local native flora Provides coastal habitat for many plants and animals Reduces erosion	Newly accreted land in Noakhali, Patuakhali, Bhola, Cox's Bazar and Bagerhat districts
	Grass turf or mattress coverage	A layer of vegetation is placed on the embankment or streambank to protect against small scale erosion from waves and wind	Increases the stability of embankments Increases stream flows Increases the aesthetic beauty of the embankment or stream	Polders and levies destroyed by cyclonic surges in Khulna, Bagerhat, Satkhira Barguna, and Patuakhali districts
	Beach, shoreline, or estuary nourishment	Sediment of suitable quality is artificially added to a beach, shoreline, or estuary that has been sediment deficient, in order to maintain stability and provide protection from tidal surge or wave action	Reduces erosion provides a storm buffer Protects the habitats and nesting grounds of aquatic species Increases the aesthetic beauty of beach, shoreline, or estuary	Kalapara (Patuakhali), and coastline along the Sundarbans in Khulna, and Bagerhat districts

(continued)

Table 5.1 (continued)

Impact area	EbA measures	Description	Ecosystem Functions and Services	Spatial coverage
Agricultural sustainability and bio-diversity conservation	Crop diversification	New cultivars or species and improved varieties are introduced, aiming at enhanced plant productivity, quality, and nutritional value, as well as greater resilience to pests, diseases and climate change	Increases productivity of food, fibre and fuel. Controls disease and pest infestation. Facilitates soil formation and nutrient cycling. Increases resilience to environmental stress	Areas affected by high salinity (Khulna, Bagerhat Satkhira), and waterlogging (Khulna, Jessore, Gopalgong districts)
	Integrated farming	A farmer optimizes production by conducting different agricultural practices together, such as crops, aquaculture, and plantation of small fruit trees	Increases food security. Maintains soil fertility. Maintains soil biodiversity. Increases nutrient cycling	Areas affected by varying level of salinity (Khulna, Bagerhat, Satkhira, Barishal)
	Agro-forestry	Agro-forestry integrates the production of trees and of non-tree crops on the same piece of land. Agro-forestry can improve the resilience of agricultural production to climate variability and change	Increases the productivity of food, fibre, and fuel. Facilitates soil formation and nutrient retention. Performs carbon sequestration for climate change mitigation. Regulates of micro-climates. Diversifies livelihoods	Interior coastal districts such as Jessore, Jhenadha, Narail, Kustia
	Ecological pest management	Natural processes of pest regulation are reinforced to improve agricultural production	reduces the production costs and environmental pollution. Enhances the natural ecosystem resilience. Increases food safety	Interior coastal district such as Jessore, Jhenadha, Magural, Kustia district

(continued)

Table 5.1 (continued)

Impact area	EbA measures	Description	Ecosystem Functions and Services	Spatial coverage
Water resource management	Coastal wetland and canal restoration	A range of activities prevents the settling of sediment load, leaching of chemicals, water pollution, degradation, as well as the overuse of coastal wetlands such as mangroves, salt marshes, and canals	provides freshwater reservoir Increases nutrient cycling in soil and water Prevents flooding and drainage congestion Provides coastal habitats for many plants and animals	Areas affected by tidal flooding and drainage congestion (Khulna, Bagerhat Satkhira, Jessore, Gopalgong district)
	Tidal river management	Water management allows sediment-borne tidal water flow into the wetland basin and releases the tidal flow back to the river. As a result of this process, sediments carried by the tidal flow are deposited on the wetland basin and improve the waterlogged condition	Increases capacity of natural drainage Decreases water logging Recovers agricultural fields from submerged conditions Improves soil condition	Areas affected by tidal flooding and drainage congestion (Khulna, Jessore district)
	Hydroponic systems (a.k.a., floating cultivation)	Floating cultivation utilizes, for food production, areas which are waterlogged for long periods of time. Floating cultivation aims to adapt to more regular or prolonged flooding	Mitigates land lost to flooding Improves livelihood and nutritional security Facilitates soil formation and nutrient recycling	Areas affected by tidal flooding and drainage congestion (Bagerhat Satkhira, Jessore, Gopalgong)
	Rainwater harvesting and storage in small reservoirs and micro-catchments	This practice consists of collecting rainfall from ground surfaces using micro-catchments to divert or slow runoff, so that it can be stored in artificial or natural courses for agricultural use and to improve soil moisture	Increases groundwater recharge Improves soil conditions Reduces erosion from surface water runoff	Areas affected by high salinity (Khulna, Bagerhat Satkhira)

EbA strategies, particularly those operated at the landscape level, should integrate multiple sectoral policies across various agencies and institutions, many of whom often have very different priorities (Ahammad et al. 2013). Different sectoral authorities often emphasize interventions in their own domain and often neglect the benefits for other sectors. For instance, to protect the coastal embankments around the Sundarbans Reserve Forest, the responsible agency, the Water Development Board, prefers engineered measures, whereas the Forest Department prefers EbA such as coastal green belts.

An accurate assessment of the sensitivity of coastal ecosystems to both climatic and non-climatic stressors is essential for EbA-centered policy. Unfortunately, systemic uncertainty hampers the timely availability of reliable data on coastal dynamics and the pragmatic design of EbA approaches. These problems render EbA less effective in building community resilience and ensuring ecosystem health (Nandy et al. 2013). Therefore, more scientific research is needed regarding the role of ecosystem services in reducing the vulnerability of coastal communities, as well as the costs and benefits of habitat conservation in the context of climate change (Ahammad et al. 2013; Alam et al. 2013). Despite these constraints, the Bangladesh NAPA and the Bangladesh Climate Change Strategy and Action Plan have recognized and incorporated scientific evidence into EbA interventions.

5.4 The Way Forward

Commonly adopted EbA measures include mangrove afforestation, coastal green belts, agro-forestry, restoration of coastal wetlands and natural canals, hydroponic agriculture, integrated farming, and other agronomic innovations. However, sustainability of many of these measures is questionable for numerous reasons, including the often top-down approaches in project implementation, the lack of in inter-agency coordination, and poor understanding about the changing dynamics of coastal systems. As EbA adoption is constrained by several institutional and policy challenges, there is a need for fresh initiatives to overcome these challenges. Accordingly, we propose the following:

- EbA should involve integrated sectoral approaches, including all relevant sectors such as forestry, fisheries, agriculture, and water resources, that account for uncertainty and risk.
- EbA policy needs a flexible social-institutional interface that would foster evolving coastal policies and institutional frameworks to anticipate climatic uncertainty.
- EbA measures should be developed in close consultation with local communities, the private sector, and other relevant stakeholders in order to incorporate local practices.
- An integrated "top-down" and "bottom-up" approach should be followed to ensure sustainable outcomes across various levels of implementation.

Box 5.1: Fisheries management in Bangladesh in the context of climate change. *Source*: Md. Monirul Islam, Department of Fisheries, University of Dhaka, Dhaka, Bangladesh

In Bangladesh, the fisheries sector contributes significantly to nutrition (provides 60% of all animal protein), employment (for 6 million fishers and fish farmers) and the economy, with fish production the source of 3.69% of GDP and 2.01% of export earnings (BDF 2015). Aquaculture, inland capture, and marine capture fisheries are all vulnerable to the effects of climate change (Islam et al. 2014a), such as increased storm intensity, floods, droughts, and temperature extremes (Met Office 2011). The consequences for fishers could range from changes in ecosystems and fish stocks (Cheung et al. 2009) to damage to equipment, property, and infrastructure (FAO 2008), and would vary across different regions and water bodies. However, few robust research studies concerning the impact of climate change on Bangladesh's fisheries exist, complicating efforts to formulate sustainable management plans.

For aquaculture, fish that tolerate temperature fluctuations, particularly high heat in the summer, will be more adaptable, since such fluctuations can significantly affect breeding and larval development. Less summer rainfall will lead to lower pond water levels, requiring species which can grow to market size under such conditions. Aquaculture producers could create shade to mitigate some temperature extremes. The development of genetically modified fish to thrive in such conditions is another possible strategy. Greater monsoon rainfall would cause pond overflow unless fish farmers raise dikes or adopt other protective strategies. Because tropical storm intensity can damage aquaculture infrastructure, the building of climate-proof infrastructure needs to be integrated into long-term development plans. However, none of these strategies are currently widely implemented in Bangladesh, except for some experimental aquaculture with temperature-tolerant tilapia (*Oreochromis niloticus*).

Increased storm intensity would negatively impact capture fisheries in rivers, estuaries, and coastal and marine areas. However, heavy rain and flooding may be beneficial to inland open water fisheries (e.g., floodplains and *haors*), by facilitating their migration, breeding, nursery and feeding. Maritime fishers are especially vulnerable to cyclones and would need to adopt changes in fishing practices and technology such as the use of modern fishing boats (Islam et al. 2014b). Ocean acidification will also negatively impact Bangladeshi marine shellfish. To respond to changes in breeding, migration, and recruitment of commercially important species, conservation strategies are needed. For example, current management of hilsa fish includes spatial and temporal capture bans, which will need to be modified as climate change disrupts breeding and migration patterns.

EbAs involve a wide range of ecosystem management activities to bolster ecosystem health and reduce vulnerability. If we can address the above issues, we can be optimistic about building resilient coastal communities despite a changing climate in Bangladesh.

References

Ahammad R, Nandy P and Husnain P (2013) Unlocking ecosystem based adaptation opportunities in coastal Bangladesh. *J Coast Conserv* 17(4):833–840.

Alam M, Ahammad R, Nandy P and Rahman S (2013) Coastal Livelihood Adaptation in Changing Climate: Bangladesh Experience of NAPA Priority Project Implementation. In: Shaw R, Mallick F, Islam A (eds) *Climate Change Adaptation Actions in Bangladesh*, Springer, Tokyo, Japan, pp. 253–276.

Ali A (1999) Climate change impacts and adaptation assessment in Bangladesh. *Climate Res* 12(2–3):109–116.

Bangladesh Department of Fisheries (BDF) (2015) *National Fish Week 2015 Compendium* (in Bangla). Department of Fisheries, Ministry of Fisheries and Livestock, Bangladesh.

Butler JRA, Wise RM, Skewes TD, Bohensky EL, Peterson N, Suadnya W, Yanuartati Y, Handayani T, Habibi P, Puspadi K, Bou N, Vaghelo D and Rochester W (2015) Integrating top-down and bottom-up adaptation planning to build adaptive capacity: a structured learning approach. *Coastal Management* 43(4):346–364.

Cheung WWL, Lam VWY, Sarmiento JL, Kearney K, Watson R and Pauly D (2009) Projecting global marine biodiversity impacts under climate change scenario. *Fish and Fisheries* 10(3):235–251.

Colls A, Ash N and Ikkala N (2009) *Ecosystem-based Adaptation: a natural response to climate change*. IUCN, Gland, Switzerland.

CSISA (2012) *Farmers manual on improved rice farming technologies*. Cereal Systems Initiative for South Asia in Bangladesh, IRRI (in Bangla).

Dickinson M, Prentice IC and Mace GM (2015) *Climate change and challenges for conservation*. Grantham Institute Briefing Paper No. 13, Imperial College London, UK.

Doswald N and Estrella M (2015) *Promoting ecosystems for disaster risk reduction and climate change adaptation: opportunities for integration*. Discussion paper, United Nations Environment Programme, Geneva.

Food and Agriculture Organization of the United Nations (FAO) (2008) *Report of the FAO expert workshop on climate change implications for fisheries and aquaculture*. Rome: Food and Agriculture Organization of the United Nations.

Government of Bangladesh (GoB) (2001) *National Water Management Plan*. Water Resources Planning Organization, Ministry of Water Resources. Government of Bangladesh.

Government of Bangladesh (GoB) (2005a) *National Adaptation Programme of Action*. Ministry of Environment and Forests. Government of Bangladesh.

Government of Bangladesh (GoB) (2005b) *Coastal Zone Policy*. Ministry of Water Resources. Government of Bangladesh.

Government of Bangladesh (GoB) (2008) *Bangladesh Climate Change Strategy and Action Plan*. Ministry of Environment and Forests. Government of Bangladesh.

Haque KNH, Chowdhury FA and Khatun KR (2014) Participatory Environmental Governance and Climate Change Adaptation: Mainstreaming of Tidal River Management in South-west Bangladesh. In: Ha Huong (ed) *Land and Disaster Management Strategies in Asia*, Springer, India. pp. 189–208.

Hossain MS, Hein L, Rip FI and Dearing JA (2015) Integrating ecosystem services and climate change responses in coastal wetlands developments plans for Bangladesh. *Mitig Adapt Strat Glob Change* 20(2):241–261.

Iftekhar MS (2006) Conservation and management of the Bangladesh coastal ecosystem: overview of an integrated approach. *Nat Res Forum* 30(3):230–237.

Iftekhar MS and Takama T (2008) Perceptions of biodiversity, environmental services, and conservation of planted mangroves: a case study on Nijhum Dwip Island, Bangladesh. *Wetlands Ecology and Management* 16(2):119–137.

IPCC (2001) *Climate change 2001: The scientific basis. Contribution of Working Group I to the Third Assessment Report of the Intergovernmental Panel on Climate Change.* Cambridge University Press, Cambridge and New York.

IPCC (2007) *Climate Change 2007: Impacts, Adaptation and Vulnerability. Contribution of Working Group II to the Fourth Assessment Report of the IPCC,* Cambridge University Press, Cambridge, England.

Islam SN (2016) Deltaic floodplains development and wetland ecosystems management in the Ganges–Brahmaputra–Meghna Rivers Delta in Bangladesh. *Sustain. Water Resour. Manag.* 2(3):237–256.

Islam MM, Sallu S, Hubacek K and Paavola J (2014a) Vulnerability of fishery-based livelihoods to the impacts of climate variability and change: insights from coastal Bangladesh. *Regional Environmental Change* 14(1):281–294.

Islam MM, Sallu S, Hubacek K and Paavola J (2014b) Limits and barriers to adaptation to climate variability and change in Bangladeshi coastal fishing communities. *Marine Policy* 43:208–216.

Islam MN and Kitazawa D (2013) Modeling of freshwater wetland management strategies for building the public awareness at local level in Bangladesh. *Mitig Adapt Strateg Glob Change* 18(6):869–888.

Islam MS and Wahab MA (2005) A review on the present status and management of mangrove wetland habitat resources in Bangladesh with emphasis on mangrove fisheries and aquaculture. *Hydrobiologia* 542(1):165–190.

IWM and CEGIS (2007). *Investigating the Impact of Relative Sea-Level Rise on Coastal Communities and their Livelihoods in Bangladesh: Final Report, June 2007.* IWM/CEGIS/ Government of Bangladesh, Dhaka, Bangladesh.

Khadka C and Vacik H (2012) Comparing a top-down and bottom-up approach in the identification of criteria and indicators for sustainable community forest management in Nepal. *Forestry* 85(1):145–158.

Mallick B, Rahman KR and Vogt J (2011) Coastal livelihood and physical infrastructure in Bangladesh after cyclone Aila. *Mitig Adapt Strateg Glob Change* 16(6):629–648.

Mamun AA (2010) Understanding the Value of Local Ecological Knowledge and Practices for Habitat Restoration in Human-Altered Floodplain Systems: A Case from Bangladesh. *Environmental Management* 45(5):922–938.

Met Office (2011). Climate: Observations, projections and impacts – Bangladesh. Met Office Hadley Centre and Department of Energy and Climate Change of the United Kingdom.

Millennium Ecosystem Assessment (MEA) (2005) *Ecosystems and human well-being: wetlands and water synthesis.* World Resources Institute, Washington, DC.

Munang R, Thiaw I, Alverson K, Mumba M, Liu J and Rivington M (2013) Climate change and Ecosystem-based Adaptation: a new pragmatic approach to buffering climate change impacts. *Current Opinion in Environmental Sustainability* 5(1):67–71.

Nandy P, Ahammad R, Alam M and Islam A (2013) Coastal Ecosystem Based Adaptation: Bangladesh Experience. In: Shaw R, Mallick F, Islam A (eds) *Climate Change Adaptation Actions in Bangladesh,* Springer, Tokyo, Japan. pp. 277–303.

Nishat A and Mukherjee N (2013). Sea Level Rise and Its Impacts in Coastal Areas of Bangladesh. In: Shaw R, Mallick F, Islam A (eds) *Climate Change Adaptation Actions in Bangladesh,* Springer Tokyo, Japan. pp. 43–50.

Paul BG and Vogl CR (2011) Impacts of shrimp farming in Bangladesh: Challenges and alternatives. *Ocean & Coastal Management,* 54(3):201–211.

PDO-ICZMP (2004). *Living in the Coast: People and Livelihoods*. Project Development Office-Integrated Coastal Zone Management Plan (PDO-ICZMP), Water Resources Planning Organization, Government of Bangladesh.

Rahman R and Salehin M (2013) Flood Risks and Reduction Approaches in Bangladesh. In: Shaw R, Mallick F, Islam A (eds) *Disaster Risk Reduction Approaches in Bangladesh*. Springer, Tokyo, Japan. pp. 65–90.

Reid H and Alam SS (2014*) Ecosystem based Approaches to Adaptation: Evidence from two sites in Bangladesh*. IIED Working Paper. IIED, London.

Saroar MM, Routray JK and Leal Filho W (2015) Livelihood vulnerability and displacement in coastal Bangladesh: Understanding the nexus. In: *Climate Change in the Asia-Pacific* Region. Springer, Berlin. pp. 9–31.

Schmitt K and Duke NC (2015) Mangrove management, assessment and monitoring. In: Pancel L, Kohl M (eds). *Tropical Forestry Handbook*. Springer-Verlag, Berlin. pp. 1725–1759.

Sultana P and Thompson PM (2007) Community Based Fisheries Management and Fisher Livelihoods: Bangladesh Case Studies. *Hum Ecol* 35(5):527–546.

Sultana P and Thompson PM (2008) Gender and local floodplain management institutions: a case study from Bangladesh. *J Int Develop* 20(1):53–68.

Uddin MS, van Steveninck ER, Stuip M and Shah MAR (2013) Economic evaluation of provisioning and cultural services of a protected mangrove ecosystem: a case study on Sundarbans Reserve Forest, Bangladesh. *Ecosyst Serv* 5:88–93.

Vignola R, Locatelli B, Martinez C and Imbach P (2009) Ecosystem-based adaptation to climate change: what role for policy makers, society and scientists? *Mitig Adapt Strateg Glob Change* 14(8):691–696.

Chapter 6
Climate Governance and Finance in Bangladesh

Mousumi Pervin, Pulak Barua, Nuzhat Imam, Md. Mahfuzul Haque and Nahrin Jannat Hossain

Abstract Bangladesh, being one of the most vulnerable countries to climate change, ratified both the United Nations Framework Convention on Climate Change (UNFCCC) and the Kyoto Protocol from the very beginning. Since the 1970s, the Government, in collaboration with development partners, has invested substantial amounts in flood protection schemes, coastal embankment projects, cyclone shelters, comprehensive disaster management projects, irrigation schemes, agricultural research and coastal 'greenbelt' projects. One of its first initiatives was the creation of a Climate Change Unit established within the Department of Environment in 2004 which provided a central focus for the Government's work on climate change. Bangladesh was the one of the few Least Developed Countries to develop a National Adaptation Programme of Action (NAPA) in 2005 in addressing countrywide programs for climate change adaptation in a holistic approach. In 2009, the Government of Bangladesh unveiled the Bangladesh Climate Change Strategy and Action Plan (BCCSAP). In its efforts to ensure adequate investment in building resilience and managing disasters, the government established a Climate Change Trust Fund in 2009 with its own money and the Bangladesh Climate Change Resilience Fund (BCCRF) in 2010 with support from development partners. Though Bangladesh has made significant accomplishments in terms of policies to address climate change compared to other developing nations, it still needs to do more to sustain growth. Therefore, it is important to translate the BCCSAP and other relevant policy recommendations into actions across institutions, scale-up the good practices at local level and use indigenous knowledge.

Mousumi Pervin, United Nations Development Programme, Bangladesh, Corresponding Author, e-mail: mousumipervin@gmail.com.

Pulak Barua, Unilever Bangladesh Ltd.

Nuzhat Imam, Christian Commission for Development in Bangladesh.

Md. Mahfuzul Haque, Transparency International Bangladesh.

Nahrin Jannat Hossain, Department of Geography and Environment, Jagannath University.

The original version of this chapter was revised: Authors' affiliations have been updated. The correction to this chapter is available at https://doi.org/10.1007/978-3-030-05237-9_14

© Springer Nature Switzerland AG 2019   65
S. Huq et al. (eds.), *Confronting Climate Change in Bangladesh*,
The Anthropocene: Politik—Economics—Society—Science 28,
https://doi.org/10.1007/978-3-030-05237-9_6

Keywords UNFCCC · Governance · Finance · Climate · Institutions Adaptation

6.1   Introduction

Development interventions have accelerated the rate of climate change, with developing and Small Island nations experiencing the worst of its adverse effects. As a result, the *United Nations Framework Convention on Climate Change* (UNFCCC) has established different mechanisms, such as *the Least Developed Countries Fund* (LDCF), *Special Climate Change Fund* (SCCF) and the Adaptation Fund, in order to disperse climate finance for adaptation and mitigation efforts to address climate change impacts. The LDCF was one of the mechanisms established in 2001 to provide support to the LDCs' climate adaptation efforts, including the preparation and implementation of *National Adaptation Programmes of Action* (NAPAs). The SCCF is the first comprehensive fund accessible to all developing countries under four funding windows: adaptation, technology transfer, sector-specific projects and assistance with diversification of fuel-dependent economies. On the other hand, the *Climate Investment Funds* (CIFs), established in 2008, were a more flexible type of financing. Implemented with the help of *the Multilateral Development Banks* (MDBs), the CIFs consist of two separate funds: the Clean Technology Fund and the *Strategic Climate Fund* (SCF). The SCF has three programs – the *Forest Investment Program* (FIP), the *Pilot Program for Climate Resilience* (PPCR) and the *Scaling Up Renewable Energy in Low Income Countries Program* (SREP) – that help the developing countries pilot low-emissions and climate resilient development. At COP15 in 2009, *First Start Finance* (FSF) was established to provide new and additional financial resources (USD30 million) to be allocated equally between mitigation and adaptation efforts from 2010–12.

In line with LDCF's priorities, Bangladesh was the one of the few Least Developed Countries who developed a National Adaptation Programme of Action (NAPA) in 2005, which was then revised in 2009. The NAPA aims to establish a holistic approach to engage in climate change adaptation at the country level. The Government of Bangladesh (GoB) also unveiled the *Bangladesh Climate Change Strategy and Action Plan* (BCCSAP) in 2008 and revised it in 2009. BCCSAP takes an integrated approach to climate change and sustainable development and focuses on poverty eradiation and well-being of the most vulnerable groups. In this chapter, we will be focusing more on adaptation finance because it defines resources and uses multilateral implementing entities of governance outside main structure of governance, whereas others are more generic in nature. There are three funding windows for channelling adaptation finance: the *Climate Change Trust Fund* (BCCTF) established in 2009, the *Bangladesh Climate Change Resilience Fund* (BCCRF) in 2010, and the PPCR which is a part of the CIFs.

The BCCTF set aside a total of USD400 million to support BCCSAP activities from 2009–2017. Of this budget, 66% has been allocated to projects, and the remaining 34% has been kept as a fixed deposit as per the Climate Change Trust

Act, 2010. BCCRF support has come in from Denmark, the European Union, Sweden and the United Kingdom. Switzerland, Australia and the United States have yet to channel in over USD188 million in grant funds to build resilience. The PPCR, designed under the leadership of the Bangladeshi government in coordination with the *Asian Development Bank* (ADB), members of the World Bank Group, key stakeholders, and other development partners, serves to channel funds and implement adaptation projects in Bangladesh. It particularly deals with additional funding to pilot innovative public and private sector solutions and it allocated USD110 million in grants (45%) and interest credits (55%).

In recent years, all major governments' national policy documents, Vision 2021 for 2010–2021, the Sixth *Five Year Plan* (FYP) (2011–2015) and the 7th FYP (2016–2020) have focused on issues around climate change. The Sixth Five Year Plan also considered climate change concerns in accelerating growth and development. Though Bangladesh has made significant progress in terms of incorporating climate change into its policies compare to other developing nations, it still needs to focus more on sustaining the growth and outgrowth of the country. Therefore, it is important to translate the BCCSAP and other relevant policy recommendations into action across institutions, scale-up the good practices at the local level and make use of the indigenous knowledge for adaptation.

The purpose of this chapter is to depict the ways in which climate change has been integrated into the planning processes in Bangladesh since the 1970s and review the effectiveness of climate finance governance. The chapter also focuses on future needs to be addressed in terms of low carbon emission development pathways and adaptation for sustainable development and inclusive growth.

6.2 Timeline of Climate Change Interventions in Bangladesh

Bangladesh, where 75% of all disasters originate from extreme climatic events, is one of the most disaster-prone countries in the world (*Ministry of Disaster Management and Relief* (MoDMR) 2010). The Bengal Famine of 1943, consecutive floods in the years 1953, 1954, 1955, a severe cyclone in 1970, prolonged floods on 1987, 1988, 1998, and 2007, a cyclone in 1991, mega cyclones Sidr in 2007 and Aila in 2009 are all known as disaster events in the history of Bangladesh. These events caused huge numbers of human causality in addition to loss and damage to assets. However, it also brought opportunities for the local people, the State and professionals to gain knowledge, experience and expertise that, over time, increased the disaster management capacity of the country (Pervin 2013). The Government, in collaboration with development partners, has invested over USD10 billion in the last two decades in different risk reduction measures – both structural and non-structural – and enhanced its disaster preparedness system (World Bank 2010). These measures have significantly reduced loss of life, injuries, and economic losses from extreme disaster events over time (Kabir/Hossain 2013).

We have divided climate change integration into planning and development into the following phases, considering the timescale and likelihood of interventions took place in Bangladesh: *(1) Initial Stage; (2) Development Stage; and (3) Advanced Stage.* The initial stage is characterized by interventions taking place, understanding and capacity being introduced to manage disasters and climate change impacts in a piecemeal manner. This stage started in the mid-1970s and continued until the start of the 21st century. During this period, the development interventions were mainly focused on relief-based approaches to disaster management. At this stage, disaster risk reduction was generally underfunded, so practitioners were likely to be on the lookout for funding sources. The development stage is characterized by specific interventions, such as the creation of policy provisions, research and technological innovations, and the establishment of funding entities. The advanced stage mainly refers to recent endeavours, i.e. from 2015 onwards, and depends upon a particular and comprehensive level of commitment on the part of the *Government of Bangladesh* (GoB).

6.2.1 Initial Stage

Since gaining independence, Bangladesh has persistently faced extreme poverty and famine, unplanned governance as well as political turmoil, which have challenged inclusive growth and development. The development interventions have focused primarily on economic development for poverty reduction in addition to a wide range of disaster risk reduction structural measures in terms of post-disaster centralized relief distribution. Over time, the development approaches have been diversified by incorporating environmental and social concerns for sustainable development in line with global responses. For example, policy concerns for environmental protection in Bangladesh were reflected for the first time in the Fourth Five Year Plan (1990–95), and have been included in other five-year plans since. This inclusion has unlocked the potential to move forward in a new direction of sustainable development, considering environmental concerns. The GoB approved the Environmental Policy and 187 statutory laws relating to environmental management (LSE 2015). Table 6.1 presents the key attributes of the stages of climate change interventions in Bangladesh.

6.2.2 Development Stage

The signing of the Kyoto Protocol by the GoB played a driving force in a paradigm shift from relief-based responses to climate and disaster risk reduction. For example, the GoB created policy provisions e.g. NAPA, NAP, BCCSAP as well as established financing mechanisms, project based interventions, sectoral integration, etc. Since the beginning of the twenty-first century, Bangladesh has made

Table 6.1 Trends of major risk reduction measures

	Key events	Major risk reduction measures	Attributes
Initial stage (1970–2000)	**1970**: Cyclone killed 300,000 people and caused USD2.5 billion worth of damage to property **1974, 1987, 1988, 1991, 1995**: Floods **1998**: Floods covered two thirds of the country, killed 1,000 people, left 30 million homeless and resulted in an economic loss of USD2 billion (4.8% of GDP)	**Structural interventions**: Costal embankments: In the 1960s, 123 polders were formed by 5017 km of embankments, of which 957 km are sea dikes Environmental safeguards measures: Mangrove and non-mangrove plantations (as of 2013, a total of 1,92,395 ha mangrove, 8,690 ha non-mangrove, 2,873 ha Nypa and 12,127 km strip plantations have been raised in the coastal areas (Hasan 2013) Shelter construction: By 2006, 2500 multi-purpose cyclone shelters had been constructed by GoB Centralized relief operation system Cyclone Preparedness Program formed in 1972 **Non-Structural Interventions**: Awareness raising about environmental management, particularly water pollution *Flood Action Plan (FAP)*: Developed to provide a permanent solution to the recurrent flood problem Creation of *Disaster Management Bureau (DMB)* in 1993, Bangladesh Meteorological Department in 1971, Bangladesh Water Development Board in 1972	**Approach** Relief-based approach to disaster management Top-down decision-making process dominated by structural measures Poverty reduction focused development interventions, but environmental safeguards measures were not foreseen **Knowledge, information and technology** Massive flood control Small scale technology: Disaster shelters, water and flood control, flood protection embankments, drainage channel excavation, sluice gates, radar for *Early Warning System* (EWS) Inadequate information management systems **State of capacity** Lack of *Operation and Maintenance* (O&M) capacity Inadequate resource accumulation and mobilization Lack of decentralized system Delayed responses, lengthy implementation processes and inadequate readiness support Lack of knowledge, capacity, experience and systems **Policy Framework** Five Year Plan (1st, 2nd and 3rd) *Standing Order for Disaster* (SOD) (1997) National Environment Policy (1992) National Forest Policy (1994), First National Energy Policy (1996) *National Environment Management Action Plan* (NEMAP) (1995–2005)

(continued)

Table 6.1 (continued)

Key events	Major risk reduction measures	Attributes
Developing stage (2001–2015)	**Structural interventions:**	**Approach**
2000, 2004: Floods	Multipurpose cyclone shelter construction	Project based approach in line with typical development
2007: Cyclone economic loss was more than USD1.7 billion (2.6% of GDP)	Relief operation: Centralized relief operation systems	Strategic natural resource management
2009: Cyclone	Afforestation and reforestation	CCA with focus on wise use of natural resources
	Livelihood diversification	**Knowledge, informaiton and technology**
	Drainage Systems and IWRM (*Integrated Water Resources Management*)	Stand-alone technological interventions
	Meteorological stations and improvements of weather forecasting and prediction	Sector specific interventions
	Renewables and solar energy	Lack of scaling up of technologies
	Non-structural interventions:	**State of capacity**
	Modeling, designing of infrastructure and early warning systems	Lack of coordination with mainstreaming development systems
	Filling in of knowledge gaps	Overlapping services
	Emergency Disaster Shelters and Information Assistance centers	Inadequate O&M capacity
	Mainstreaming adaptation to CC into policies and programs	Lack of coordination between mainstreaming development systems and climate specific interventions
	Awareness and behavioral change and communication	**Policy framework**
	Capacity building and knowledge management	*Poverty Reduction Strategy Paper (PRSP)* I and II
	Promotion of research	Five Year Plan (4th, 5th)
	Strengthening of human resource capacity and institutions	Perspective Plan
	Establishment of Climate Funds (BCCTF and BCCRF)	NAPA (2005, updated 2009)
	Resilience building	Bangladesh Climate Change Strategy and Action Plan (2008, revised 2009)
		Renewable Energy Policy (2009)
		National Sustainable Development Strategy Revised SOD (2010)
		National Plan for Disaster Management (NPDM) (2010–15)
		BCCTF Act (2010)
		New clause (18A) on environment in Constitution in 2011
		National Sustainable Development Strategy (2013)
		National Environment Policy (2013, published in draft form, updating the previous 1992 policy)

(continued)

Table 6.1 (continued)

	Key events	Major risk reduction measures	Attributes
Advanced (2015 onwards)	Sea level rise (SLR) Severity and intensity of climate induced disasters Environmental degradation New development challenges Public health concerns	**Structural interventions:** Designing resilient infrastructure Strengthening early warning systems and dissemination mechanisms **Non-structural interventions:** Institutional strengthening Cross-coordination among institutions and sectors Sustainability of interventions Community based adaptation and Ecosystem based adaptation Loss and Damage *Sustainable Development Goals* (SDGs)	**Approach** Inclusiveness through climate resilient and low carbon emission pathways Cross-cutting integration with mainstreaming development Paradigm shift from relief to resilience development **Knowledge, information and technology** Energy efficient technology for low carbon emissions Scaling up of technologies Access to information through digitalization of services REDD+, low emission development pathways, green growth **State of capacity** Widespread solar power (mainly solar home systems and solar irrigation; grid-connected solar power has not started yet, though it's in the pipeline) Strong policy platforms but a poor enabling environment Lack of synergies among policies and institutions Lack of innovation of technologies and limited investment to research and study **Policy framework** *National Adaptation Plan (NAP)* initiated *Nationally Appropriate Mitigation Action (NAMA)* initiated Bangladesh Delta Action Plan initiated Five Year Plan (7th) (2016–2020) INDC 2015, Climate Fiscal Framework (2015) Climate Fiscal Framework DM Policy (2015)

Source ???

Table 6.2 Number, amount and major activities of projects in three different funds

Fund name		Number of projects	Amount in USD million	
			Allocation in funds	Approved in projects
BCCTF	Govt. and technical assistance projects	377	400	298.5
	NGO	63		
BCCRF	Government	12	186.8	157.3
	NGO	38		
PPCR		6	671.8	99.3
Total		425	1258.6	555.1

Source Compiled by the author from BCCTF (2014), BCCRF (2014), and CIF (2015a, b)

significant accomplishments in terms of policy perspectives and piloting innovative adaptation options in addressing climate change. The knowledge capacity in the country, compared to many other LDCs is relatively high and policies and institutions are taking shape.

6.2.2.1 Broad Based Integration into Central Planning

The most comprehensive approach has been followed by the government since 2000 with the help of central planning documents like the sixth (1997–2002) and seventh (2011–2015) Five Year Plans. The Poverty Reduction Strategy Paper linked development activities with the environment, with a focus towards improving the quality of life. Even the constitution of Bangladesh (18A) in 2011 introduced a new clause for safeguarding and developing the environment and wildlife. Moreover, the GoB developed the 'Outline Perspective Plan (OPP) of Bangladesh 2010–2021: Making Vision 2021 a Reality' which includes a dedicated chapter on the environment and disaster risk management, where vulnerabilities, responses and climate change management strategies are analysed.

6.2.2.2 Climate Change Exclusive Planning

Along with the central planning process, climate change exclusive planning exists where CC specific interventions are identified and budgeted for. For example, in 2005, the GoB became the first LDC to develop a NAPA, which identified 15 priority adaptation programs and activities. It became evident that the NAPA was insufficient to tackle the dramatic impacts faced by Bangladesh. This led to the revision of the NAPA in 2009, and the development of the *Bangladesh Climate Change Strategy and Action Plan* (BCCSAP), which aims to integrate climate change constraints and opportunities into overall planning and programs, involving all sectors and processes for economic and social development. As Bangladesh is at the forefront of climate change, the GoB shared its plans for BCCSAP with development partners at the London Climate Summit in September 2008. At this summit, development partners pledged financing to implement BCCSAP (Alam et al. 2010). These pledges enabled BCCSAP to take effect in 2009.

BCCSAP takes an integrated approach to climate change and sustainable development and focuses on poverty eradication and well-being of the most vulnerable groups. In its efforts to ensure adequate investment in building resilience and managing disasters, the GoB established the Climate Change Trust Fund in 2009 with its own money and the *Bangladesh Climate Change Resilience Fund* (BCCRF) in 2010 with support from development partners. The BCCRF began operating in May 2010.

The *Pilot Program for Climate Resilience* (PPCR) of the *Strategic Climate Fund* (SCF), was established under the Multi-donor *Climate Investment Funds* (CIF) with the aim to bring transformational change in climate adaptation. In the first phase of PPCR, a *strategic program for climate resilience* (SPCR) was approved by the government in 2010. The second phase of PPCR, which is being led by the Asian Development Bank with technical support from the World Bank and the International Finance Corporation, focuses on implementing the SPCR through actions such as support to policy reform, institutional capacity building, and scaling-up other investments in key sectors.

PPCR provided additional funding to put the plan into action and pilot innovative public and private sector solutions to pressing climate-related risks. According to the World Bank, the PPCR programmatic approach entails a long-term, strategic arrangement of linked investment projects and activities to achieve large-scale, systematic impacts and take advantage of synergies and co–financing opportunities. Moreover, it provides grants and highly concessional financing for investments supporting a wide range of activities. The GoB received grants and concessionary loans worth USD110 million (USD50 million grant and USD60 million near-zero interest concessional loan) from PPCR, based on the SPCR documents. There has also been some co-financing from the *International Fund for Agricultural Development* (IFAD), *Bill and Melinda Gates Foundation* (BMGF), Asian Development Bank (ADB), World Bank and the *Government of Bangladesh* (GoB).

6.2.3   Advanced Stage

It is evident that the GoB has already created provisions for addressing climate change issues and it is now time for remedial actions for future implementation because environmental degradation will be exacerbated by frequent disasters and climate change, making the situation worse. Moreover, the population will stabilize at around 200 million and growing industrialization, displacement and forced migration due to environmental challenges will place further enormous strains on ecosystems and the living environment.

Since 2015, all policy documents have focused on broad-based integration rather than narrowed, isolated interventions. They cover mainstreaming, technological interventions, green growth, mitigation and adaptation, resilience development, loss and damage, and regional cooperation for sustainable development. Loss and damage is a relative newcomer to the climate change agenda. 'Loss' is characterized as the negative impacts of climate change that are permanent, and 'damage' as those impacts that can be reversed (Huq et al. 2013).

The perspective plan emphasizes climate change adaptation through the active participation of local communities and the private sector, rather than a top-down strategy. Within the upcoming Seventh FYP (2016–20), articulation of a sustainable development strategy involves three key themes: (i) CC Management and Resilience (comprised of adaptation and mitigation); (ii) Environmental Management; and (iii) Disaster Management. The actions under these themes are aligned with the overall framework and strategies of the *National Sustainable Development Strategy 2010–21* (NSDS), and are broadly consistent with the scope of the post-2015 *Sustainable Development Goals* (SDGs).

Recently, Bangladesh and The Netherlands have been working together to formulate a long-term plan for management of the Bangladesh delta – known as the *Bangladesh Delta Plan* (BDP) 2100 – with the General Economics Division of the Planning Commission. The BDP 2100 aims to integrate planning, both spatially and sectorally, and to frame strategy making in relation to trends, scenarios and a long-term vision, thus connecting it to the central planning documents and other existing sectoral plans, policies and strategies. It will contribute to disaster risk reduction, water safety and CC resilience and adaptation, food security and economic development in Bangladesh.

The Paris Agreement was signed by 175 countries at COP21 in Paris in December 2015. Bangladesh outlined and submitted its *Intended Nationally Determined Contributions* (INDCs) proposal in September 2015 to reduce future emissions as part of a robust and ambitious international agreement. It sets emissions reduction goals for the power, transport, and industry sectors, alongside further mitigation actions (USD27 billion for mitigation and USD42 billion for adaptation from 2015–2030) in other sectors and intends to implement its conditional emissions reduction goal subject to appropriate international support in the form of finance, investment, technology development and transfer, and capacity building.

Bangladesh is currently finalizing its National Adaptation Plan (NAP) roadmap, with financial support from the Government of Norway and technical assistance from the NAP Global Support Program. UNEP and UNDP are supporting Bangladesh on adaptation with financing from the *Global Environment Facility's* (GEF) Least Developed Countries Fund. Additional initiatives include the UNEP/UNDP *Poverty-Environment Initiative* (PEI) and the PPCR, alongside those from the *Food and Agriculture Organization* (FAO), *International Fund for Agricultural Development* (IFAD) and *Asian Development Bank* (ADB). The PPCR received funding from the CIF. DFID, GIZ and KFW are also providing support for addressing climate change.

6.3   The Role of Civil Society in Climate Change Planning in Bangladesh

Civil society refers to the wide array of non-governmental and not-for-profit organizations that have a presence in public life, expressing the interests and values of their members or others, based on ethical, cultural, political, scientific, religious or philanthropic considerations (World Bank 2013). In Bangladesh, civil society has a long history of contributing to poverty alleviation efforts. This tradition has extended into the area of adaptation, facilitated by programs such as *Action Research for Community Adaptation in Bangladesh* (ARCAB) and 'Gobeshona' (Bengali term for research and is a knowledge sharing platform for climate change research on Bangladesh). While there is a wealth of information available and both the GoB and many *civil society organizations* (CSOs) are taking initiative in the areas of climate change, effective adaptation is constrained by limited exchange of information between and within the GoB and civil society organizations (Thomalla et al. 2005). Over the last few decades, civil society organizations (CSOs) have played an active role in the areas of economic development, poverty alleviation efforts and human rights, but they are yet to establish themselves as important actors in the areas of environmental conservation and climate change planning (Rahman 1999; Islam 1999). However, there are a few exceptions – civil society organizations have played an important role in strengthening the capabilities in the areas of disaster management (Thomalla et al. 2005). The role of CSOs is still at an early stage in Bangladesh due to a variety of reasons. Firstly, most of the environmental CSOs are largely donor funded by foreign resource-dependent NGOs who define the project and, in many cases, have specific advocacy agendas of their own which may not directly link to climate change planning. Secondly, in Bangladesh, most of the CSOs are fragmented with their approach and activities and do not have effective networking among themselves and with public sector institutions, resulting in less effective exchanges of knowledge and a lack of influence. Thirdly, environmental and climate change planning and adaptation have yet to enter the agenda of all sections of Bangladesh's civil society, ranging from professional and trade organizations to literary and cultural organizations who can play an important role in advocacy and public awareness.

There are some exceptions where CSOs have provided significant input and contribution towards developing some policy documents such as the NAPA, BCCSAP, etc. In its executive summary, Bangladesh's NAPA claims that different civil society representatives, including academic and research institutions, doctors, media, *non-governmental organizations* (NGOs) and *Community Based Organization* (CBO) representatives contributed to the development of the NAPA.

A number of citizen-based environmental campaigns, e.g. the *Bangladesh Paribesh Andolon* (BAPA) and *Paribesh Bachao Andolon* (POBA), have contributed significantly to the development of different government policy documents related to CC. Some other civil society groups, e.g. the Equity and Justice Working Group (Equity), which is a coalition of national NGOs, and the Oxfam-led *campaign for Sustainable Rural Livelihoods* (CSRL), have played an important role in the BCCSAP process as well. Having networks with global climate justice campaigns, they mobilize significant public opinion around climate change issues by organizing national and international events (Alam et al. 2013).

CSOs have also been trying to gain a firm footing in bringing transparency and clarity to climate finance and governance. Transparency International appears to be very active in making transparency, accountability and oversight key to ensuring climate finance funds are free from corruption. To fight with the impacts of climate change and to develop the necessary adaptation capability, Bangladesh needs to have proper planning as well as huge financial and technological support from international communities. Hence, climate finance and proper governance also become critical to the agenda for Bangladesh, where a well-functioning civil society is a fundamental pre-condition to achieving success.

6.4 The Geography of Governance and Finance

6.4.1 Climate Finance Governance in Bangladesh

There are diverse intermediaries, instruments and planning systems in Bangladesh's financial landscape (Pervin/Islam 2014). The Ministry of Environment and Forests (MoEF) is the leading institute responsible for managing climate finance as part of its activities to promote the BCCSAP. MoEF operates two funds: BCCTF and BCCRF. In order to ensure good management of these funds, the Bangladesh Climate Change Trust Act was established in 2010. MoEF also established a separate unit, known as the *Bangladesh Climate Change Trust* (BCCT), to administrate the fund (see Fig. 6.1). *The Pilot Program for Climate Resilience* (PPCR), a funding window of the *Climate Investment Fund* (CIF), is helping developing countries integrate climate resilience into development planning and offers additional funding to support public and private sector investments for implementation. Though the funds are similar in many ways, the BCCRF and PPCR fund fewer projects than the BCCTF, but those projects that they do fund are bigger and more comprehensive.

The BCCRF and PPCR are also efficient in resource management. They have internal monitoring and reporting mechanisms and disclosure policies, which are absent from the BCCTF. The difference between the BCCRF and PPCR is that the contribution to the BCCRF comes as pure grants while the contributions in PPCR are a combination of both grants and credits.

Fig. 6.1 Governance
structure of climate finance.
Source Authors own

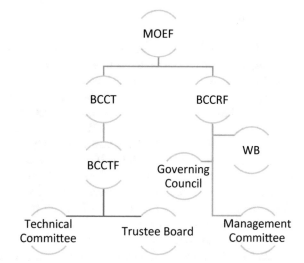

6.4.1.1 BCCTF

Established with revenue, the BCCTF is a block budgetary allocation in the form of an endowment by the Government of Bangladesh. As per the Climate Change Trust Act 2010, a maximum of 66% of the allocated amount, as well as the interests accrued on the remaining 34% (which is kept as fixed deposit), can be allocated to CCTF projects. The remaining 34% of the fund is kept for use in case of an emergency. Since 2009–10, a total of BDT31 billion (approx. USD400 million) has been allocated to the BCCTF. As of June 2016, 440 projects had been undertaken. 377 projects are being implemented by government, semi-government and autonomous agencies, while 63 projects are being implemented by NGOs under the *Palli Karma Sahayak Foundation* (PKSF). Eight projects have been cancelled due to misconduct. Thus far, BDT250 million (approx. USD3.19 million) has been allocated to the PKSF for NGO projects. Among them, a total 129 (110 governments and 43 NGO) projects have already been completed (BCCT 2014).

The BCCTF is governed by a board of trustees, technical committee and sub-technical committees. The board of trustees is chaired by the MoEF and is responsible for granting approval of projects, guiding the overall strategy and activities of the fund, and reviewing progress. A technical committee, chaired by the Secretary of the MoEF, is charged with undertaking thorough technical reviews of submitted proposals and screening them for recommendation to the Governing Council.

6.4.1.2 BCCRF

BCCRF is a multi-donor trust fund created to channel climate funds from development partners by employing the World Bank as fiduciary manager. The fund was established in May 2010 to build resilience to the effects of CC. According to BCCRF's (2014) annual report, including the supplemental contributions by Sweden and Switzerland in 2012, the total amount pledged to the BCCRF as of the end of 2014 was approximately USD187 million. At the end of 2014, five investment projects and five analytical and advisory activities, totalling USD86.6 million had been approved, and USD34.3 million had been disbursed. No supplemental financial contribution has been made since then.

BCCRF's governance structure consists of three tiers: (i) a *Governing Council* (GC); (ii) a *Management Committee* (MC); and (iii) a Secretariat. The Governing Council (GC) provides overall strategic direction and guidance to BCCRF and ensures its alignment with the BCCSAP. It is a high-level committee chaired by the MoEF. The *Management Committee* (MC) is a small technical committee chaired by the Secretary of MoEF. It is responsible for the work program, ensuring that the BCCRF is implemented in line with the agreed Implementation Manual. On February 23, 2011, the MC approved the establishment of a BCCRF secretariat at MoEF to support the administration of BCCRF activities. The GC subsequently approved an allocation of USD0.2 million on May 19, 2011 for the establishment of the secretariat. In practice, the secretariat was not functioning as anticipated throughout 2014 and, consequently, a request for additional allocation of funds to support further capacity building was not approved.

6.4.1.3 PPCR

PPCR is a funding window from Climate Investment Funds (CIF) to assist developing countries in integrating climate resilience into development planning. Bangladesh's PPCR investment plan was designed under the leadership of the government, in coordination with the Asian Development Bank (ADB), members of the World Bank Group, International Bank for Reconstruction and Development (IBRD), International Development Association (IDA), International Finance Corporation (IFC), key Bangladeshi stakeholders, and other development partners.

6.4.2 Status of Funds Execution

A total of 425 climate finance projects in Bangladesh are at different stages of implementation under the three funds.

6.4.3 Geography of Climate Change Finance Investment

BCCTF projects have been broadly implemented all over the country, with a particular focus on the construction, repair, and maintenance of infrastructure for river banks, polders, embankments and drainage systems under municipalities (Table 6.3 and Fig. 6.2). Out of 306 government projects, 47 projects, worth a total of USD31.6 million, have been approved for the Chittagong district, and 18 projects, worth a total of USD15.7 million, are being implemented in the Bhola districts, while parts of an additional 12 projects have been approved for different locations in Bhola. Furthermore, for the district Pirojpur, 14 projects and components of another eight projects, worth USD13.9 million, have been sanctioned in favour of food security, social protection and infrastructure related activities. Other coastal districts such as Patuakhali, Madaripur, Satkhira, Barisal, Barguna, Khulna and Cox's Bazar have received a significant amount of money from BCCTF (Table 6.2). However, BCCTF projects have focused on the Chittagong, Cox's Bazar and Bhola districts (Fig. 6.2) for the construction of coastal sea dykes, embankments, drainage systems, and water control infrastructure including regulators and sluice gates. Re-excavation of canals and river bank protective works have also been done in Barisal, Madaripur, Faridpur, Gopalganj and Satkhira districts. Comprehensive disaster management projects, particularly the construction of

Table 6.3 Major activities under climate funds in Bangladesh

BCCTF	BCCRF	PPCR
Construction, repair and maintenance of dykes, cyclone resilient houses, embankments, drainage, water sources, regulators Protection and re-excavation of canals and rivers Development of early warning and weather forecasting systems, stress tolerant seeds, floating agriculture systems Tree plantation and re-vegetation Distribution of biogas plants, solar system and improved cooking strove (BCCTF 2014)	Improving climate resilience of coastal populations to tropical cyclones Improving the capacity of the *Ministry of Environment and Forests* (MoEF) to manage donor-funded climate change activities Enhancing the capacity of selected communities to increase their resilience to the impacts of climate change Reducing forest degradation and increasing forest coverage Increasing access to clean energy in rural areas through renewable energy and promoting more efficient energy consumption (BCCRF 2014)	Improving climate resilient agriculture and food security Strengthening the security and reliability of fresh water supply, sanitation, and infrastructure, and enhancing the resilience of coastal communities and infrastructure Technical assistance aimed at enabling Bangladesh to better assess and address the effects of climate change to support capacity development and knowledge management initiatives Coastal housing feasibility study (CIF 2015a, b)

Source BCCTF (2014) and CIF (2015a, b)

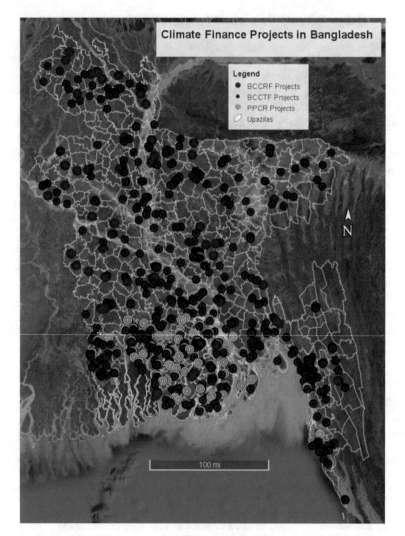

Fig. 6.2 Climate finance projects in Bangladesh (BCCRF, BCCTF and PPCR). *Source* The Authors

cyclone resilient houses, have been implemented in Khulna, Satkhira and Bagerhat districts (BCCTF 2014).

BCCRF projects have been approved for adaptation, mitigation, low carbon development, and infrastructural activities and most of these projects are being implemented in the Chittagong, Cox's Bazar, Barguna, Pirojpur, Khulna, Satkhira and Patuakhali districts. In addition, 38 small NGO projects from Community Climate Change Projects (CCCP) have been approved for coastal, drought and flood prone areas of Bangladesh. PPCR projects are exclusively focusing on adaptation activities in 12 polders in the coastal districts of Satkhira, Khulna, Bagerhat, Pirojpur, Barguna, Patuakhali, Bhola, Noakhali, Lakshmipur, Feni, Chittagong and Cox's Bazar (Fig. 6.2) (CIF 2010). The Chittagong has received funding for the most projects (50) from the three funds, while Patuakhali received the second highest (funding for 31 projects), with Pirojpur not far behind with 27 projects. 58 municipalities of different districts also got projects funded from the BCCTF, though the municipality projects given in some municipalities of Mymensingh, Bogra, Gazipur, Rajbari, Rangpur, Brahmanbaria are not taking place in very high-risk zones of climate change. This raises the question of whether risk and vulnerability-based priorities were considered when determining where such projects would occur.

6.5   Conclusions

Climate change has a disproportionately stronger impact on the lives and livelihoods of poor people, but there are still many challenges associated with addressing these climate change effects in Bangladesh. Over the last couple of decades, climate change impacts and consecutive disastrous events have changed the spatial distributions and vulnerabilities in Bangladesh, particularly in the coastal and drought prone regions. Therefore, considering the changing conditions, it is important to review policies because these documents are based on relatively old information, and communities should be involved in this revision process. There are existing policies, strategies and plans that have taken into consideration present problems and growing trends, but they are not flexible enough to accommodate the longer-term effects of climate change. Therefore, a long term strategic plan in line with the existing one could be instrumental to the long-term sustainability of development efforts. Existing efforts should build upon synergies instead of creating parallel and uncoordinated systems. The coordination process may follow steps like (1) aligning the efforts with priorities of other funding windows; (2) capitalizing funds; (3) assembling governance bodies; (4) establishing horizontal and vertical feedback systems; (5) creating a standard fiduciary system; and (6) facilitating a central knowledge hub for information sharing. In addition, a strong participatory monitoring, evaluation and disclosure mechanism would help measure the effectiveness of climate finance.

In the future, global climate fund disbursement or availability will not be based solely on vulnerability, but rather mostly on a country's ability to ensure effectiveness of the utilization of funds. There is an opportunity for Bangladesh to strengthen its capacity and set up systems to prove the efficiency to spend climate funds effectively in order to access future climate funds such as the Green Climate Fund and other global climate funds. Bangladesh uses a range of financial intermediaries to access climate funds from public and private sources. There is a need to scale-up efforts to access international funds and innovative blending of different financial mechanisms. Moreover, effective use of public funds must be ensured to leverage private financing through an equitable sharing of responsibilities among different stakeholders.

Climate change is of a cross-cutting nature and there is a risk of potential consequences of decisions in one sector affecting another sector. As a result, policy-makers and decision-makers across all areas and levels of government, as well as in the private sector and civil society, need to be aware of the importance of considering disaster risk and climate change and incorporate it into their work. Otherwise, it will be difficult to maintain the appropriate management mechanisms and commitment to climate change adaptation and mitigation will decrease, having long-term consequences for Bangladesh.

References

Alam, K., Shamsuddoha, M., Tanner, T., Sultana, M., Huq, M., and Kabir, S. (2010). *Planning exceptionalism? Political Economy of Climate Resilient Development in Bangladesh.* Institute of Development Studies; at: https://www.ids.ac.uk/files/dmfile/BangladeshPECCMainReportFinal2. pdf (27 Sept 2015).

BCCRF (2014). *Bangladesh Climate Change Resilience Fund. Annual Report 2013.*

BCCT (2014). *Project Profile 2009–10, 2010–11, 2011–12, 2012–13, 2013–2014.* Ministry of Environment and Forest. Government of the People's Republic of Bangladesh. Dhaka, Bangladesh. Print copies.

BCCTF (2014). Bangladesh Climate Change Trust (BCCT). Ministry of Environment and Forests; at: http://www.bcct.gov.bd/index.php/83-home/147-home-2 (27 Sept 2015).

CIF (2010). Bangladesh: Strategic Program for Climate Resilience (SPCR).

CIF (2015a). *Climate Investment Funds.*

CIF (2015b). *PPCR investment plan. Strategic Program for Climate Resilience (SPCR).*

Huq, S., Roberts, E., and Fenton, A. (2013). Loss and Damage, *Nature Climate Change,* Vol 3, Macmillan Publishers Limited, November 2013.

Hasan, D. Z. (2013). *Plants in Mangroves and Coastal Afforestation in Bangladesh.* Dewan House, Ukilpara, Naogaon-6500, Bangladesh, pp. 164.

Islam, N. (1999). Protecting Bangladesh's environment: the role of the civil society. *Journal of Social Studies* (Dhaka): 34–63.

Kabir, M., and Hossain, M. (2013). *Climate Change and Sustainable Development: Bangladesh Perspective.* Conference Paper, Conference: International conference on Climate Change Impact and Adaptation, At DUET-Gazipur, Bangladesh, November 2013; at: http://www. researchgate.net/publication/275100243_CLIMATE_CHANGE_AND_SUSTAINABLE_ DEVELOPMENT_BANGLADESH_PERSPECTIVE (30 November, 2015).

MoDMR (2010). *National Plan for Disaster Management: 2010–2015,* Ministry of Disaster Management and Relief, Government of the People's Republic of Bangladesh, Dhaka, Bangladesh.

Pervin, M. (2013). *Mainstreaming climate change resilience into development planning in Bangladesh: A country Report,* IIED, UK.

Pervin, M., and Islam, M. (2014). *Climate finance governance in Bangladesh: synergies in the financial landscape, A Policy Brief*, IIED, UK.

Rahman, A. (1999) 'NGOs and civil society in Bangladesh'. *Journal of Social Studies*, 84: 23–45.

Thomalla, F., Cannon, T., Huq, S., Klein, R. J., and Schaerer, C. (2005, May). Mainstreaming adaptation to climate change in coastal Bangladesh by building civil society alliances. In *Proceedings of the solutions to coastal disasters conference* (pp. 668–684).

Chapter 7
The Role of State Broadcasting Media and Education in Addressing Climate Change in Bangladesh

Afifa Afroz, Safayet Khan, Ishrat Binte Mahmud and Mohammad Nazmul Chowdhury

Abstract Media and education undoubtedly play an important role in alerting and preparing people from many kinds of natural calamities and disasters. This is especially true for disasters that are caused by climate change. Through media and education people not only become more aware of the changes happening in nature, but are also empowered to minimise the risks associated with it. This chapter discusses several strategies in the domain of media and education to effectively tackle climate change impacts. It includes approaches taken by multiple stakeholders, including the Bangladesh Government and different non-government organisations. The Bangladesh Government's preferred invisibilist approach of knowledge dissemination versus different community organisations' supported visibilist approach of local knowledge integration has been discussed. Different forms of public media and, most importantly, state broadcast media are currently engaged in this process. Educational interventions coordinated by government and non-government organisations are also addressing the issue. This chapter provides examples of different forms of communication interventions that can assist in creating and raising public awareness to combat the impacts of climate change in Bangladesh. However, the combined roles of media and education in addressing climate change and community wellbeing need to be investigated further in future impact evaluation studies.

Keywords Media · Broadcasting · Education · Communities

Afifa Afroz, Bangladesh Television, Television Bhaban, Rampura, Dhaka, Corresponding Author, e-mail: afifa_afroz@yahoo.com.

Safayet Khan, BRAC Research and Evaluation Division, BRAC Centre, 75 Mohakhali, Dhaka-1212.

Ishrat Binte Mahmud, British Standard School; 9B Gopi Kishan Lane, Wari, Dhaka-1203.

Mohammad Nazmul Chowdhury, ICCCAD; House 27, Road 1, Block A, Bashundhara R/A, Dhaka 1220.

© Springer Nature Switzerland AG 2019   85
S. Huq et al. (eds.), *Confronting Climate Change in Bangladesh*,
The Anthropocene: Politik—Economics—Society—Science 28,
https://doi.org/10.1007/978-3-030-05237-9_7

7.1   Introduction

The geographical location of Bangladesh makes the country susceptible to natural disasters. Climate change has increased this susceptibility (MoFDM 2007), further complicating the already precarious socio-economic and demographic conditions of the country. Current and future generations will suffer from increasingly intense climatic events throughout their lives.

Climatic events such as cyclones and their associated storm surges, droughts, floods, waterlogging, saline water intrusion, river erosion etc., are experienced in many parts of Bangladesh. The coastal areas of Bangladesh are regularly hit by tropical cyclones, which cause loss of valuable lives and property. Statistics indicate a probability of 1.12 cyclones hitting Bangladesh in any given year (Sarker/Azam 2007). Low-lying depressed basins (*Haors*) in northeastern Bangladesh experience flash floods. These damage crops and cause a decline in fish populations (Anik/Khan 2012). In the same areas, soil erosion in the monsoon period damages infrastructure, causing the break-down of roads and embankments. Prolonged drought is a common phenomenon in northwestern parts of the country and cause even greater damage to crops than flooding and submergence (World Bank 2013b).

Those people impacted by climatic disasters are taking action to adapt to the changing climate patterns. Adaptive activities and measures include, but are not limited to, early warning systems, multipurpose cyclone shelter construction, cluster housing patterns and house fortifications, resilient plantations in coastal areas, livelihood diversification and agricultural extension services to improve agricultural practices (Sarker/Azam 2007). These are multi-stakeholder efforts designed to build resilience among vulnerable populations.

The need to raise public awareness and understanding of climatic risks through disseminating accurate information is important to building long-term resilience to climate change (UNISDR 2015). Such understanding is crucial to establishing and developing innovative adaptive and responsive measures, such as those indicated above. Yet, climate change and its long-term multiple effects are not always directly or easily perceivable. External descriptions of climate risks are primarily produced by scientists, and often communicated in a complex way. This makes it difficult for the general public to understand these risks. Instead, audience-specific messaging and framing for active engagement (Moser 2010) between multiple stakeholders can better support understanding. In addition to this, knowledge of local communities and their adaptation techniques needs to be integrated to the overall communication and education programmes for better uptake of the delivered messages within communities.

The role of media and education for building climate and disaster resilient communities is recognised at the global level. For example, the role of media has been advocated by initiatives such as the Sendai Framework for Disaster Risk Reduction (2015–2030). This framework emphasises the utilisation and strengthening of all kinds of media, including traditional media such as television, radio, newspapers, magazines, newsletters, tax press and other print publications (Zhao et al. 2011; UNISDR 2015), to support successful communication of disaster risk.

The role of education is key to a holistic response to climate change, enhancing knowledge and offering innovative ideas to increase resilience and build security from local to national and international levels (UNFCCC 2007; UNISDR 2015). Recognition of *Climate Change Education* (CCE) is growing and, in recent years, has developed its own identity through UNESCO's campaign for Education on Sustainable Development[1] (UNESCO 2006; Læssøe et al. 2009). In Bangladesh, communication of climate change knowledge can help prepare vulnerable people and enable them to respond to related challenges. However, it is crucial that information is disseminated in a simple, transparent and accessible manner, by engaging media and other stakeholders, including the authorities at all levels.

This chapter discusses several strategies and provides examples of different forms of communication interventions through media and education, which help to create public awareness. Although different types of media are mentioned, particular emphasis has been given to state broadcasting media, as they are the main recipients of initial warnings and government standing orders to ensure rapid transmission to the general public. However, other private print and electronic media also play an important role in transmitting important messages (Habib et al. 2012). This chapter aims to assess different initiatives taken by the government, non-government and community-based organisations. It will ground these initiatives in a theoretical framework surrounding how people perceive climate change in order to determine the primary means of communicating and disseminating knowledge of climate change in Bangladesh.

7.2 Communicating Perceptions and Understandings of Climate Change

There are two predominant schools of thought on human perceptibility of climate change. The *invisibilist* approach considers perceptibility of climate change to be dependent on the dissemination of climatic information via scientific, technical or institutional networks (Mormont/Dasnoy 1995). On the other hand, the *visibilist* approach contributes importance to the reporting of climate change by local people who are experiencing its effects (Riedlinger/Berkes 2001; Green et al. 2010). A new emerging school of thought, known as the *constructive visibilism* approach, argues that reports on climate change should reflect both local peoples' knowledge and external information on climate change (Marin/Berkes 2013). Importantly, the receiver of external information must perceive the information to be credible in order for them to pay it any attention. Perceived credibility is dependent on the source of the information. In a bureaucratic and hierarchical society such as Bangladesh, it is observed that the higher the authority of the official source

---

[1]UNSESCO led a decade long campaign from 2005 to 2014 that aimed to develop locally relevant high quality, holistic Education on Sustainable Development to foster critical thinking and problem-solving (UNESCO 2006).

(e.g. that from government ministries or departments), the greater its appeal (Allan 1999; Denham 2010). This consideration must be built into communication approaches in order for information to impact upon its receiver.

There are various approaches in operation to communicate perceptive and external climate change preparedness knowledge. Media is the main source of information for lay people as well as decision-makers (Arlt et al. 2011; Stamm et al. 2000). Over time, media has shaped people's perceptions and understanding of climate change-related disasters and impacted the scientific and policy discourse at national and international levels. Education is also used to enhance knowledge and incite innovative ideas that increase resilience within communities (Bonifacio et al. 2010; UNISDR 2005). It can provide people with practical knowledge for immediate and long-term responses to climate change, for example by sharing adaptation knowledge for livelihood improvement, or by sharing mitigation knowledge to mitigate the impacts of disasters. While these different types of knowledge can prepare society to deal with the challenges posed by climate-induced disasters, without effective and functional communication systems, information cannot be translated into action and the knowledge on these critical issues could be lost (Olausson/Berglez 2014). Communication mechanisms such as media and education need to complement one another to provide a comprehensive system for reciprocally communicating knowledge between the public domain, the scientific arena and other pools of knowledge from alternative stakeholder groups.

In Bangladesh, however, little research has been done to examine the range of adaptation measures using different types of education and media. The Bangladesh education system is complex (World Bank 2013a). Research on different adaptation measures appropriate for different educational contexts such as general education, madrasah education, English-medium education and vocational or technical education has not been done yet. This is also true for different types of media, such as print and electronic (audio and audio-visual) media. Individual experiences and existing foreknowledge play an important role in human learning processes stated in the moderate constructivism learning theory (Kattmann 2003). Therefore, knowledge transfer via media and knowledge-sharing through a certain degree of educational instruction is needed for reconstruction of concepts in the target group (Riemeier 2007; Riede et al. 2017). Research is needed in this domain to examine the combined role of media and education and its influence on policy decisions regarding curriculum development and knowledge dissemination. A developed curriculum can ensure that students at all levels are provided with technical knowledge of adaptation and mitigation options. On the other hand, dissemination of knowledge to the mass population can be achieved by effectively using different forms of media – e.g. through "edutainment"[2] (Rey-López et al. 2006). How these types of interventions achieve broader positive impacts on community wellbeing in Bangladesh needs to be examined.

---

[2]"Edutainment" refers to educational content that has entertainment value or vice versa. This form of media has been used by the governments in different countries since 1970. It is used to disseminate information on health and social issues to impact on viewers' opinions and behaviour (Rosin 2006).

7.3   Media and Climate Change

7.3.1   Media Operations

Media are means of communication that distribute content – such as text, pictures and sound – to an anonymous and spatially diverse public via technical means (McQuail 2010). There are both traditional and new types of media. Traditional media includes both printed and electronic media, such as newspapers, magazines or books, and broadcast media, such as radio, television or film. The new media includes publicly accessible websites such as online newspapers, blogs, wikis, video games, and social media (Klöckner 2015). Media outlets and different stakeholder organisations (e.g. political parties, non-governmental organisations (NGOs), private companies and social media operatives) form a different segment of media.

7.3.2   The Use of Media in Bangladesh

The uses of media in Bangladesh are varied and consist of both traditional and new media. However, traditional media is more visible than new media, as very few people have internet access in Bangladesh. The total number of internet subscribers reached 67.2 million in 2017 and this number is increasing as more people start to use internet through mobile phones (BTRC 2017b). On the other hand, traditional media includes both print and electronic media, which consist of both state and privately-owned audio and audio-visual media. *Bangladesh Sangbad Sanngstha* (BSS) is the country's national news agency. In addition, 2 state-owned broadcasting media channels (Bangladesh Television and Bangladesh Betar), 26 private satellite television channels, 25 FM broadcasting channels and 16 community radio channels are in operation in Bangladesh (BTRC 2017a). Although different types of media have been mentioned, particular emphasis has been given to state broadcasting media such as BSS, Bangladesh Television and Bangladesh Betar. These are the main recipients of initial warnings and government standing orders for rapid transmission to the general public. However, other private print and electronic media also play an important role in this regard.

Radio and television are two central media types that enable communication of information to the masses. Radio has been considered a primary medium through which information is communicated to the mass public in Bangladesh. However, ownership of a radio in Bangladesh has halved, from 8 percent in 2011 to 4 percent in 2014 (NIPORT 2015). This is reflected in listening rates. According to the Bangladesh Demographic and Health Survey of 2011, the proportion of women who listen to the radio every week has declined sharply from 19 percent in 2009 to only 5 percent in 2011. Similarly, the proportion of men who regularly listen to radio has almost quartered, from 38 percent in 2007 to only 10 percent in 2011 (NIPORT 2013). In contrast, TV ownership has increased significantly over the last 20 years

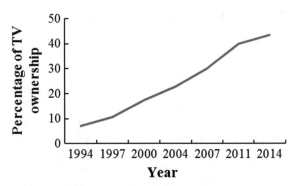

Fig. 7.1 Increase of TV ownership over 20 years in Bangladesh. *Source* NIPORT (2013)

(see Fig. 7.1). By 2014, 43.5 percent of the population of Bangladesh owned a television (NIPORT 2015), and 52 percent (82.4 million people) were estimated to watch television (BBC 2014). The decline in the use of radios might have been a result of this increased access to and use of televisions, mobile telephones and the use of internet in those phones. The total number of mobile phone subscription in Bangladesh reached around 122 million by the end of January 2015 (BTRC 2017c).

As an audio-visual type of media, television has great appeal. It enables viewers to understand information regardless of their level of literacy. Thus, television can influence a much wider audience. In this respect, state public broadcasters such as *Bangladesh Television* (BTV) and Bangladesh Betar, can play a significant role in raising public awareness on critical issues. BTV is the only terrestrial TV Channel in Bangladesh. As a state-owned television channel, it has a greater reach than other private satellite channels. A media survey conducted by BBC Media Action in 2014 found that out of the 82.4 million people who watch television, 49.4 million (60 percent) watch BTV and 14.32 million (17 percent) watch BTV exclusively (BBC 2014). Figure 7.2 indicates the broadcasting rates of different categories of television programmes broadcasted on BTV between 2009 and 2014. This provides an example of the composition of different programmes usually aired by state broadcasting authority. This further corroborates the commitment of a state broadcasting agency towards building of a learning society. The figure shows that news, education and informative programmes dominate the BTV broadcasting schedule, with 52 percent of programmes fitting into these categories.

7.3.3 Media Communication for Climatic Responsiveness

The media framing of climate change combines policy, science and public perspectives. It is important for influencing audiences and stimulating their subsequent motivation to take action (Swain 2012). For most adults in Bangladesh,[3] media, and particularly local television news programmes, are among the major sources of

---

[3]Aged 21 years and above.

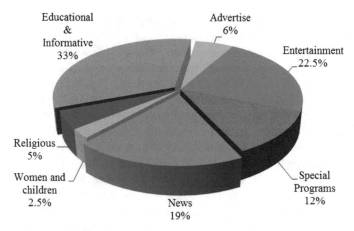

Fig. 7.2 Distribution of BTV programmes from 2009 to 2014 (The data for this figure is based on hourly distributions of programmes collected from programme cue sheet and TV guides). *Source* Authors using data from BTV

environmental information, providing regular weather updates through special weather bulletins. A BBC Media Action study found that 93 percent of the audience of BTV regularly watch "*Raat At tar Sangbad*" (News at Eight) (BBC 2014). Due to the link between news and environmental information provision, this viewing preference has significant implications for enhancing viewers' awareness of environmental issues (Coyle 2005; Kohut/Remez 2008). BTV could provide an important source of information on climate change, natural disasters and hazards for its viewers, helping to build awareness on different types of disasters in the community. Other private television channels are also playing an important role in this regard.

The role of broadcasting channels in disseminating information in pre-, post- and during disaster periods has proven to be vital for community preparedness to disaster risks. In addition to news shows, both BTV and Bangladesh Betar telecast special education and awareness programmes related to disaster risk reduction and climate change that help to build the resilience of communities before disasters occur. These are produced in coordination with the Bangladesh Meteorological Department, Disaster Management Bureau and Ministry of Disaster Management and Relief. They include daily broadcasts of weather news and year-round government disaster notices, instructions and standing orders to reduce the risks of natural disasters. During disaster periods, hourly forecasting and warning news enables people to respond to rapidly changing situations. In the run up to cyclonic events, BTV and Bangladesh Betar telecast precautionary signals provided by the Meteorological Department. They explain the meaning of these signals to viewers. BTV also telecast special weather bulletins that provide audio briefings from the cyclone forecaster of the Storm Warning Centre (SWC) and narrated video representations of radar and satellite pictures. They telecast Danger Signals every 30 min, increasing to every 15 minutes when there is a "Great Danger Signal".[4] They continue these signals beyond normal broadcasting hours.

Given resource constraints and the lack of infrastructural facilities in Bangladesh, the cyclone warning dissemination process and pre-disaster preparedness through BTV and Bangladesh Betar are hugely valuable. They have been successful in alerting people to potential threats and recommending emergency measures for natural disasters. Historically, cyclones have left Bangladesh with devastating loss. Cyclones in 1970 left between 300,000 and 500,000 people dead and in 1991 a cyclone took 138,000 lives. The casualty caused by cyclones in earlier times was high due to the lack of coordinated warning dissemination to the people living in the coastal regions. Since the nineties many types of media have played a significant role in alerting the public and creating mass awareness on disaster related issues. Public awareness campaigns have been provided by different media groups (MRDI 2010). Print media reports mainly assisted on-ground efforts by different stakeholders during and after the cyclone. Although special issues on climate change, editorials and roundtable discussions with experts are regularly published, the coverage of reports on climate change are insufficient in Bangladeshi print media (Haque et al. 2010). A report published by UNDP in 2008 states that 260 press releases on disaster risk reduction and climate change were published across all print media of Bangladesh (NAARI 2010). However, print media is utilised to provide widespread news coverage before and after major cyclones (MRDI 2010). The result of these different media efforts was exemplified in 2007 when Cyclone Sidr struck Bangladesh. Five days before Sidr made landfall, the Bangladesh Government began to broadcast warnings on radio and television and it issued emergency evacuation orders almost twenty-seven hours before landfall. This helped a successful evacuation of coastal residents from the cyclone's projected path (Paul/Dutt 2010; Paul 2009). As a result of this and other different factors such as cyclone preparedness programmes, cyclone warning systems, public cyclone shelters etc. (Paul 2009), the death toll was less than 4000, a huge reduction from the hundreds of thousands of lives lost in previous years.

7.4   Education and Climate Change

7.4.1   The Relationship between Climate Change and Education

There is a two-way relationship between climate change and education. Climatic impacts, particularly disasters, have a direct and physical impact on education systems (GoB 2008). They damage school buildings and severely affect the lives and

---

[4]This is one of the 11 types of cyclone warning signals used in Bangladesh. Different classifications of weather signal are issued based upon the increasing intensity of the storm. The "Great Danger Signal" is issued when the severity of the storm exceeds the level of the "Danger Signal VII".

livelihoods of students and teachers alike. An estimated 5,927 educational institutes were fully or partly damaged by Cyclone Sidr. As a result, school children could not complete their education on time. Their physical and mental growth was also affected and many children suffered from diarrhoea, dysentery and malnutrition (Kabir et al. 2016; GoB 2008). These hinder the educational progress being sought in Bangladesh as part of the country's overarching development objectives. However, at the same time, education can support and promote climate change preparedness and response. It can be used as a tool to raise awareness and understanding of climate change and provide tangible solutions to climatic challenges. In Bangladesh, formal and informal educational activities for citizens of all ages are being adopted to combat climatic impacts. Key stakeholders, including communities, governments, development organisations and media, are working together in educational contexts to apply innovative solutions. For example, the Ministry of Education is undertaking a joint development programme in collaboration with the Ministry of Food and Disaster Management, and other relevant ministries and agencies, to establish academic building-cum-cyclone shelters in the high-risk zones of the coastal areas (GoB 2008).

7.4.2   *Responding to Climatic Impacts on Education*

School systems in Bangladesh are detrimentally impacted by climate change and need to be adapted to withstand climatic changes. There are several initiatives underway to integrate provisions for education into adaptation methods. For example, in cyclone prone areas, cyclone shelters are being used as schools during the periods that they are not in use. This helps to maintain the shelters while also providing a space for educating local students. Monsoon rains submerge many localities in the Northwestern and Northeastern parts of Bangladesh for around six months each year. In these flood prone areas, floating education systems have been introduced (see Case Study 1). These enable students to continue their education despite significant flooding and long-term water logging, which have previously disrupted the use of educational and other facilities.

Case Study 1: Floating Schools

Parts of Bangladesh are vulnerable to seasonal flash floods, which are intensifying as a result of climate change. Such climatic phenomena are disrupting children's ability to attend school, particularly for those who live in remote places such as Char and Haor areas. NGOs working in these areas are responding to this by converting traditional boats into schools, libraries and training centres (Yee 2013). The boats have open spaces that can cater for the needs of 30 children. They have waterproof roofs, equipped with solar panels, which make the boats self-sufficient. These "Floating Schools" are used to collect children from their homes. They are docked at suitable destinations while teachers conduct on-board classes. At the end of the month, student progress is assessed through examinations, followed by open progressive discussions. The floating education system also includes late evening classes for working children and women. The teachers are from the local community and they are trained to facilitate learning using different types of publications, including national curriculum books. Tutorials, training materials, presentations and books are based upon community needs, customised to different target groups, including literate and illiterate children, men and women. Some NGOs also provide wireless internet on-board the boats, enabling young people to learn and practice their computer and internet skills. The floating library is also used by adults in the community, who have meetings on global warming, rising sea levels and the impacts of climate change on farming communities. Educational activities have provided farmers with adaptation ideas and techniques. For example, farmers have implemented floating gardens as a result of the floating education systems, which in turn contribute to the health and nutrition of the community. This educational system also integrates media communication approaches, providing the opportunity for people to watch documentaries on adaptive cropping, poultry and fishing methods on-board the boat. These types of initiatives have helped thousands of families to overcome the educational obstacles that they are exposed to due to their residential location.

7.4.3   Using Education to Combat Climate Change

Climate Change Education (CCE) can reduce the climatic vulnerability of school children (Anderson 2012). Children living in poverty, especially in low-income countries with weak governance and poor education system are among those most affected by climate change. Disasters induced by climate change can destroy school facilities. CCE delivered in schools and learning centres provide education on immediate and long-term responses and on community-level sustainable development (Anderson 2012). This can include adaptation knowledge, which is instantly usable for livelihood improvement, and mitigation knowledge geared towards lifestyle changes for reduced carbon footprint, which can help to reduce the future challenges faced by society. Children get 83 percent of their environmental information from the media (Stahl et al. 1996). However, they are likely to rely on the first information they encounter and ignore subsequent conflicting information. This is due to an absence of instruction on how to critically analyse and question information received. In this instance, CCE has a role both in providing its audience with reliable and usable information and in developing students' analytical skills to enable critical thinking about information from various sources (Vaughter 2016). Moreover, linking students' knowledge to action on climate change will be more successful if educational institutes have operational policies that allow students to practice action competence. This would help new behaviours persist as they are practiced enough to become habits (Knussen/Yule 2008).

In Bangladesh, an official CCE curriculum is supported by specific books on the topic that aim to raise student awareness and, ultimately, contribute to the reduction of local and national vulnerability. The Government's National Education Policy 2010 includes a series of books, entitled "Bangladesh and Global Studies" for secondary and lower secondary school curricula. These were published in 2012, and adopted as compulsory textbooks for Classes 3 to 10 by the National Curriculum and Textbook Board. The books are printed in Bengali and English and distributed free of cost to all students. They provide comprehensive information on the science of global warming and climate change, the occurrence of natural hazards in Bangladesh and disaster risk reduction in the national and global context, with chapters focusing on both impacts and responses. The books take a forward-looking perspective to prepare learners for the future. The content is balanced and relevant, taking an interdisciplinary approach to CCE.

NGOs and development partners have also created books to support CCE. For example, *"Bangladesher Prakritic Durjog"* (The Natural Calamities of Bangladesh) was published by the Deutsche *Gesellschaft fur International Zusammenarbeit* (GIZ) GmbH and the *International Union for Conservation of Nature* (IUCN) in 2014. The use of the book is being piloted in the coastal regions of Bangladesh, under the *Coastal Livelihood Adaptation Project* (CLAP) (GIZ/IUCN 2014). This book targets primary and secondary school children. It provides engaging examples of adaptation and mitigation exercises, hazard mapping, early warning systems and disaster preparedness activities in the local context through creative approaches, such as drama and eco-activities, to provide readers with an alternative insight into issues of climate change and disasters. The handbook, *"Jolbaiyu Poribartan Ovijozon Bishoyok Handbook"* (Climate Change Adaptation Related Handbook), was published by Save the Children and Uddipon in 2014, under the *Integrated Children Centric Climate Change Adaptation Project* (ICCCCAP). It provides a comprehensive overview of adaptation techniques in rural Bangladesh and real-life lessons from the field (Uddipon 2014). The handbook focuses on creative solutions to enhance the knowledge of school children, providing customised solutions to the impacts of climate change on members of the community. Such books represent tangible progress in addressing climate change through education, and a starting point for introducing increasingly effective and wide reaching CCE approaches.

Education at the university level can complement CCE at primary, secondary and higher secondary levels. In 1992, a total of eight public universities offered environmental-related courses under different departments. However, since 1995, environmental studies in the higher education system curricula has developed significantly (Masum/Akhir 2010). Some public and private universities in Bangladesh such as the Institute of Disaster Management and Vulnerabilities Studies at Dhaka University, *Centre for Climate Change and Environment Research* (C3ER) at BRAC University etc., now have centres for CCE and offer relevant master's programmes. The curriculum contents are comprehensive, and material is delivered to the learners through intensive and extensive training sessions (Masum/Akhir 2010).

Youth are the decision-makers of tomorrow. In addition to the formal education activities described, informal education activities can be used to engage youth in every sphere of the society. Bangladesh can benefit from its young demographic (Matin 2012) to achieve a wider societal reach, greater public understanding of climate change and sustainable coping solutions. Case Study 2 highlights the contribution of youth action to educating and creating mass awareness on climate change and its associated disasters in Bangladesh.

Case Study 2: The Climate Youth Initiative

Bangladesh has made considerable progress in meeting the targets of the *Millennium Development Goals* (MDGs), however climatic disasters hinder the country's development. According to the Department of Youth Development of the Government of Bangladesh, one third of the total population in Bangladesh is aged 13–35 and therefore falls into the 'youth' category. This important portion of the population can be of great strength to Bangladesh. The British Council (2010) reports that 98 percent of young people think that they should be involved in social work and 95 percent of young people are willing to address local and community issues. However, the problem that stands is that the majority of the youth (94 percent) cannot identify an organisation or movement that specifically focuses on their generation's needs (British Council 2010).

In response to this, the *International Centre for Climate Change and Development* (ICCCAD) has established a Climate Youth Initiative Programme. The objective of this programme is to spread climate knowledge and awareness through media and educational institutions. The youth of Bangladesh are a central stakeholder in this programme. They engage in knowledge exchange and capacity-building activities to identify and put into action efforts that address the pressing needs and challenges faced by their communities.

A platform has been established for young people to discuss climate change topics or issues among themselves and prepare to present the results of their discussions to the experts at different levels. Within this platform, discussions among young professionals, activists, university students and other interested young people are organised to gain their perspectives on climate change and potential solutions. The programme then connects youth with different stakeholders such as different government departments and NGOs, so that their voices can be heard. In addition, capacity-building activities, such as training and mentoring on efficient climate-response practices enable young people to convert their knowledge into realistic actions. For example, the youth can offer their services as volunteers or interns within various organisations. The various forms of multi-stakeholder engagement sought through the Climate Youth Initiative Programme elicits support from respective stakeholder groups.

Government support for this generation could maximise the strength and creativity of Bangladesh's youth, enabling them to have greater positive impact.

7.5 Discussions and Conclusions

Bangladesh is taking a holistic approach towards combating climate change. The Bangladesh Government follow the invisibilist approach to climate change perceptibility (Mormont/Dasnoy 1995), which emphasises that information should be disseminated via institutional networks, so that the authenticity of the information is ensured. They emphasise the need to build strategic partnerships that enable different stakeholder groups in the domain of media and education to work together to provide climate-related information to the mass population. To support this, the Government of Bangladesh has formulated the National Broadcasting Policy 2014, which makes a provision for telecasting emergency weather bulletins and producing climate change awareness programmes (Chap. 3; Article 3.2; Clause 3.2.3) (GoB 2014). However, effective communication can be challenged by limitations in communication within different institutions, which is needed in order to successfully implement the Government Standing Orders. The strategic institutional framework for government ministries to interact on climate change issues is not clearly articulated in the *Bangladesh Climate Change Strategic Action Plan* (BCCSAP) (Molla 2016). Moreover, the information provided is not always grounded in the reality of citizen experience. The visibilist approach is supported by other stakeholders, such as community based organisations and NGOs working on community-based adaptation. This gives credit to the knowledge of people living with, and frequently impacted by, climate-induced disasters. This includes a wealth of traditional knowledge that has been preserved for generations alongside locally adaptive mechanisms for survival. These knowledge sources and reporting of climate change by the local people are being integrated within media and education-related interventions to appropriately support effective climate adaptive resilience within communities.

While the public need to have access to accurate information regarding climate change and disasters, it is equally important that they have the required level of ability to enable their comprehension and assimilation of disseminated information into their lives, in order to make positive and timely decisions. This is where multi-stakeholder education can support action. Bangladesh's National Education Policy 2010 emphasises the need for climate change education and the need to build a clear understanding of nature and socio-environment related matters among the public (Chap. 1; aims and objectives, 18) (GoB 2010). Educational institutions and communities across the country should be engaged to find gaps between existing educational policies and practices and be assisted to develop and mobilise national models of exemplary mitigation and adaptation approaches for the communities. Besides these, existing monitoring and evaluation mechanisms of the relevant implementing organisations should be strengthened to track progress of any interventions that create resilience within local communities by creating awareness on adaptation and mitigation. A long-term impact evaluation study can also be designed to generate evidence that can later direct and influence related policy decisions.

Different types of communications interventions, implemented via media and education activities, can promote behavioural changes needed to enhance an effective response to climate change. Since the causes of climate change are partly linked to human actions, learning to change consumption patterns using renewable forms of energy and designing greener technologies, transforms the general public into conscious consumers and responsible citizens (Anderson 2012). These modes of communications have critical roles to play in redefining peoples' lifestyles and can raise awareness of climate change issues among the mass public. These also promote long-term sustainability of interventions and build the capacity of a population to respond to environmental and developmental challenges (UNCED 1992). Such activities directly benefit people who are particularly vulnerable to disasters associated with climate change because of their locations. Therefore, people should be made aware about the particular environment in which they live and work. Current media activities, including frequent radio, TV programmes and publication in local, national, and international newspapers and magazines are beneficial in targeting those who have received a formal education as well as those who have not (Cuc 2014). Articles written in daily newspapers reach a wide range of audiences and can support and supplement radio and TV programmes. School and university systems and curricula further support awareness-raising and response. Schools can go further to promote "science literacy" (Nelson 1999) to enable students to make educated decisions about information they receive on climate change, DRR and preparedness. Within all of these awareness activities, the knowledge of local communities and their adaptation techniques needs to be integrated to support the contextual relevance of communication and education programmes.

Acknowledgment The authors would like to thank *Gobeshona* for selecting the abstracts submitted by the authors for writing a complete book chapter.

References

Allan S (1999) *News culture*. Cambridge Univ Press.
Anderson A (2012) Climate change education for mitigation and adaptation. *Journal of Education for Sustainable Development* 6 (2):191–206.

Anik SI, Khan MASA (2012) Climate change adaptation through local knowledge in the north eastern region of Bangladesh. *Mitigation and Adaptation Strategies for Global Change* 17 (8):879–896.

Arlt D, Hoppe I, Wolling J (2011) Climate change and media usage: Effects on problem awareness and behavioural intentions. *International Communication Gazette* 73 (1–2):45–63.

BBC (2014) *Know your audiences: Understanding BTV audience.* Unpublished report.

British Council (2010*) Bangladesh: The next generation.* British Council, Dhaka, Bangladesh.

Bonifacio AC, Takeuchi Y, Shaw R (2010) *Mainstreaming climate change adaptation and disaster risk reduction through school education: Perspectives and challenges.* Emerald Group Publishing, Bingley, UK.

BTRC (2017a) Bangladesh Telecommunication Regulatory Commission, Broadcasting.

BTRC (2017b) Bangladesh Telecommunication Regulatory Commission, Internet Subscription Rate.

BTRC (2017c) Bangladesh Telecommunication Regulatory Commission Mobile Phone Subscription Rate. 2017.

Coyle K (2005) *Environmental literacy in America: What ten years of NEETF/Roper research and related studies say about environmental literacy in the US.* National Environmental Education & Training Foundation.

Cuc MC (2014) The influence of media on formal and informal Education. *Procedia-Social and Behavioral Sciences* 143:68–72.

Denham BE (2010) Toward conceptual consistency in studies of agenda-building processes: A scholarly review. *The Review of Communication* 10 (4):306–323.

GIZ, IUCN (2014) *Bangladesher Prakritic Durjog* (Natural Calamities of Bangladesh). GIZ & IUCN, Dhaka.

GoB (2008) *Cyclone Sidr in Bangladesh: Damage, Loss and Needs Assessment for Disaster Recovery and Reconstruction.* Economic Relations Division, Ministry of Finance, Government of Bangladesh.

GoB (2010) *National Education Policy 2010.* Bangladesh Government, Dhaka, Bangladesh.

GoB (2014) *Bangladesh Gazette.* Bangladesh Government Press, Dhaka, Bangladesh.

Gough S, Scott W (2008) *Higher education and sustainable development: Paradox and possibility.* Routledge.

Green D, Billy J, Tapim A (2010) Indigenous Australians' knowledge of weather and climate. *Climatic Change* 100 (2):337–354.

Habib A, Shahidullah M, Ahmed D (2012) The Bangladesh cyclone preparedness program. A vital component of the nation's multi-hazard early warning system. In: *Institutional Partnerships in Multi-Hazard Early Warning Systems.* Springer, pp 29–62.

Haque MA, Yamamoto S, Sauerborn R Print Media and Climate Change in Bangladesh: The Missing Health Issue. In: *Proc. of International Conference on Environmental Aspects of Bangladesh (ICEAB10), Japan, 2010.* Institute of Public Health, Heidelberg University, Germany.

Kabir R, Khan HT, Ball E, Caldwell K (2016) Climate change impact: The experience of the coastal areas of Bangladesh affected by Cyclones Sidr and Aila. *Journal of Environmental and Public Health* 2016.

Kattmann U (2003) Scientific Literacy und kumulatives Lernen im Biologieunterricht und darüber hinaus-Ein Beitrag zur Bildungsdiskussion nach PISA 2000. Entwicklung von Wissen und Kompetenzen Internationale Tagung der Sektion Biologiedidaktik im VDBiol:99–103.

Klöckner CA (2015) Traditional and new media—About amplification and negation. *In: The Psychology of Pro-Environmental Communication.* Springer, pp 119–145.

Knussen C, Yule F (2008) "I'm not in the habit of recycling": The role of habitual behavior in the disposal of household waste. *Environment and Behavior* 40 (5):683–702.

Kohut A, Remez M (2008) *Internet overtakes newspapers as news outlet.* Pew Research Centre.

Læssøe J, Schnack K, Breiting S, Rolls S, Feinstein N, Goh KC (2009) *Climate change and sustainable development: The response from education. A Cross-National Report*, Denmark: International Alliance of Leading Education Institutes.

Marin A, Berkes F (2013) Local people's accounts of climate change: To what extent are they influenced by the media? *Wiley Interdisciplinary Reviews: Climate Change* 4 (1):1–8.

Masum A, Akhir M (2010) Environmental Education in Bangladesh with Special Reference to Higher Studies (1992–2009). *International Journal of History & Research* 1 (1).

Matin KA (2012) A Demographic Divident in Bangladesh: An Illustrative Study. In: *Presentation at the 18th Biennial Conference of the Bangladesh Economic Association to be held on, 2012.* pp 12–14.

McQuail D (2010) *McQuail's mass communication theory.* Sage Publications.

MoFDM (2007) *Draft National Plan for Disaster Management 2007–2015.* Government of the People's Republic of Bangladesh, Dhaka.

Molla IH (2016) Environment and Climate Change in Bangladesh: Challenges and the Role of Public Administration. *Resources and Environment* 6 (1):1–8.

Mormont M, Dasnoy C (1995) Source strategies and the mediatization of climate-change. Media, *Culture & Society* 17 (1).

Moser SC (2010) Communicating climate change: history, challenges, process and future directions. *Wiley Interdisciplinary Reviews: Climate Change* 1 (1):31–53.

MRDI (2010) *Needs assessment study on media capacity building in disaster reporting.* Management and Resources Development Initiative, Dhaka, Bangladesh.

NAARI (2010) *Media toolkit on disaster risk management in Bangladesh.* DIPECHO Partners, Dhaka, Bangladesh.

Nelson GD (1999) Science Literacy for All in the 21st Century. *Educational Leadership* 57 (2):14–17.

NIPORT (2013) *Bangladesh Demographic and Health Survey 2011.* NIPORT, Mitra and Associates, and ICF International., Dhaka, Bangladesh and Calverton, Maryland, USA.

NIPORT (2015) *Bangladesh Demographic and Health Survey 2014: Key Indicators.* NIPORT, Mitra and Associates, and ICF International, Dhaka, Bangladesh and Rockville, Maryland, USA.

Olausson U, Berglez P (2014) Media and climate change: Four long-standing research challenges revisited. *Environmental Communication* 8 (2):249–265.

Paul BK (2009) Why relatively fewer people died? The case of Bangladesh's Cyclone Sidr. *Natural Hazards* 50 (2):289–304.

Paul BK, Dutt S (2010) Hazard warnings and responses to evacuation orders: the case of Bangladesh's cyclone Sidr. *Geographical Review* 100 (3):336–355.

Rey-López M, Fernández-Vilas A, Díaz-Redondo RP (2006) A model for personalized learning through IDTV. In: *Adaptive Hypermedia and Adaptive Web-Based Systems, 2006.* Springer, pp 457–461.

Riede M, Keller L, Oberrauch A, Link S (2017) Climate change communication beyond the 'ivory tower': A case study about the development, application and evaluation of a science-education approach to communicate climate change to young people. *Journal of Sustainability Education* 12.

Riedlinger D, Berkes F (2001) Contributions of traditional knowledge to understanding climate change in the Canadian Arctic. *Polar Record* 37 (203):315–328.

Riemeier T (2007) *Moderater Konstruktivismus. Theorien in der biologiedidaktischen Forschung*:69–79.

Rosin H (2006) *Life lessons.* The New Yorker:40–45.

Sarker T, Azam M (2007) Super Cyclone SIDR 2007: climate change adaptation mechanisms for coastal communities in Bangladesh. Sarker, T and Azam, M (2012) Super cyclone SIDR: 85–105.

Stahl SA, Hynd CR, Britton BK, McNish MM, Bosquet D (1996) What happens when students read multiple source documents in history? *Reading Research Quarterly* 31 (4):430–456.

Stamm KR, Clark F, Eblacas PR (2000) Mass communication and public understanding of environmental problems: The case of global warming. *Public Understanding of Science* 9 (3):219–237.

Swain KA (2012) Mass Media Roles in Climate Change Mitigation. In: *Handbook of Climate Change Mitigation.* Springer, pp 161–195.

Uddipon (2014) *Jolbaiyu Poribartan Ovijozon Bishoyok Handbook* (A Handbook on Climate Change Adaptation). Integrated Children Centric Climate Change Adaptation Project (ICCCCAP), Dhaka, Bangladesh.

UNCED (1992) *The United Nations Programme of Action from Rio*: *Agenda 21*. United Nations, New York.

UNESCO (2006) *Framework for the UNDESD international implementation scheme.* ED/DESD/2006/PI/1). Paris, FR: UNESCO.

UNFCCC Impacts, vulnerabilities and adaptation in developing countries. In: *UN FCCC-United Nations Framework Convention on Climate Change*. Bonn, 2007.

UNISDR Hyogo framework for action 2005–2015: building the resilience of nations and communities to disasters. In: *Extract from the final report of the World Conference on Disaster Reduction* (A/CONF. 206/6), 2005.

UNISDR (2015) *Sendai Framework for Disaster Risk Reduction 2015–2030*. The United Nations Office for Disaster Risk Reduction, Geneva, Switzerland.

Vaughter P (2016) *Climate Change Education: From Critical Thinking to Critical Action*. UNU-IAS Policy Brief Series. United Nations, Tokyo.

World Bank (2013a) *Bangladesh Education Sector Review. Seeding fertile ground: Education that works for Bangladesh*. Human Development Report, South Asia Region. World Bank, Washington D.C.

World Bank (2013b) *Turn down the heat: climate extremes, regional impacts, and the case for resilience. A report for the World Bank by the Potsdam Institute for climate impact research and climate analytics*. World Bank, Washington D.C.

Yee A (2013) Floating Schools' Bring Classrooms to Stranded Students. *The New York Times* (July 1, 2013).

Zhao WX, Jiang J, Weng J, He J, Lim E-P, Yan H, Li X Comparing twitter and traditional media using topic models. In: *European Conference on Information Retrieval*, 2011. Springer, pp 338–349.

Chapter 8
Climate Change Is *Not* Gender Neutral: Gender Inequality, Rights and Vulnerabilities in Bangladesh

Amy Reggers

Abstract Impacts of, and responses to, climate change are not gender neutral. Climate change affects women and men differently. However, nature itself is not discriminatory. It is the social norms and gender inequalities in society that determine the differentiated impacts of climate change on women and men. Unequal power relations, both formal, such as within institutions, and informal, such as within communities and the private sphere, are at the root of the disproportionate vulnerability of women compared to men (Resurreccion et al. 2014). Yet, most gender and climate change research to date has focused on women and their specific vulnerabilities (Otzelberger 2011), rather than focusing on the ways in which inequalities contribute to vulnerabilities and hence, gender relations contribute to the differentiated effects of climate change on women and men. In Bangladesh, gender inequalities dictate that women are more affected by climate change than men. This chapter highlights a few of these underlying gender inequalities from a perspective of rights. The chapter begins with a boxed text outlining international gender and climate change policy, followed by a section on the Bangladesh context and specific gender inequalities and discriminations that contribute to Bangladeshi women's vulnerability. The author provides examples of national and community level efforts to address the gender dimensions of climate change, and demonstrates that gender stereotypes and the traditional roles of women underpin much of these efforts. Yet the opportunity is now for the Government of Bangladesh, development partners, civil society and academia to bring the issues of discrimination and rights, as they relate to gender equality, to the forefront in addressing the human side climate change impacts and consequently contribute to more gender equality across the country.

Keywords Gender inequality · Climate change · Vulnerabilities
Bangladesh

Amy Reggers, Technical Consultant, UNWomen, House # 11A #113 2, CES (C) 23, A), Road 118, Dhaka 1212, Bangladesh, Corresponding Author, e-mail: amylreggers@gmail.com.

© Springer Nature Switzerland AG 2019   103
S. Huq et al. (eds.), *Confronting Climate Change in Bangladesh*,
The Anthropocene: Politik—Economics—Society—Science 28,
https://doi.org/10.1007/978-3-030-05237-9_8

8.1 Introduction

> Nature is squarely blamed, and taming nature is then presented as the only solution.
> ("Control measures" like embankments and irrigation reflect that paradigm, and have
> therefore been common, despite their frequent counter-productiveness.) However, if natural
> forces alone were to blame, then disasters would have had an equal impact on all people.
> (Oxfam 2008: 6)

Although Bangladesh currently ranks 146 out of 187 countries in the *Human
Development Index* (HDI), it has achieved enormous progress, having doubled its
HDI score since 1980 (United Nations Development Programme 2014).
Bangladesh has decreased its poverty rate from 56.6% in 1991 to 31.5% in 2013;
however more than 40 million people still live below the poverty line (United
Nations Development Programme 2014). Coupling this, Bangladesh is widely
recognized as one of the countries most vulnerable to climate change and disasters.
The *Climate Change Vulnerability Index* (CCVI) evaluates the sensitivity of populations, the physical exposure of countries, and governmental capacity to adapt to
climate change. Among the 32 'most at risk' countries, the CCVI 2015 identifies
Bangladesh as number one (Maplecroft 2015). Despite the relatively small size of
the country, annual hazards such as floods, droughts and cyclones impact different
corners of Bangladesh, affecting large numbers of people, often those living in
poverty, in perilous conditions in rural and remote areas. Within these climate
vulnerable populations, women are recognized as one of the most vulnerable groups
in society.

Box 8.1: Gender in International Climate Change Policy by Fahmida Akhter,
Mafizur Rahman.

It is widely recognized that climate change does not affect people equally.
The related disasters and impacts often intensify existing inequalities, vulnerabilities, economic poverty and unequal power relations (Brody et al.
2008; IPCC 2007). Differently positioned women and men perceive and
experience climate change in diverse ways because of their distinct socially
constructed gender roles, responsibilities, status and identities, which result in
varied coping strategies and responses (Lambrou and Nelson 2010; FAO
2010a). Nevertheless, the issue of gender equality and women's participation
get little attention in the climate change policy regimes. The limited participation or, in some cases, complete exclusion of women from climate change
decision-making processes present a real challenge to women's empowerment, fail to uphold human rights principles and deprives society of many
skills, experiences and capacities unique to women (UNDP 2009). At COP13
(2007) in Bali, *GenderCC* – the Women for Climate Justice network, a
platform for information, knowledge and networking on gender and climate
change – was established with a coalition of women's organizations and
individuals, as well as the Global Gender and Climate Change Alliance of UN

organizations, IUCN *(International Union for Nature Conservation)* and WEDO *(Women's Environment and Development Organization). GenderCC* is actively participating in UNFCCC conferences and advocating for upholding women's rights in climate change policies. Moreover, the Hyogo Framework for Action, which emerged from the *World Conference on Disaster Reduction* (2005), states that a gender perspective should be integrated into all disaster risk management policies, plans and decision-making processes, including those related to risk assessment, early warning, information management, and education and training. The International Conference on Population and Development (1994), the Beijing Declaration and Platform for Action (1995), the World Summit on Sustainable Development (2002), and the 2005 World Summit have all recognized the essential role that women play in sustainable development. In its recent follow-up to the Beijing Platform for Action (2005), the General Assembly highlighted the need to involve women actively in environmental decision-making at all levels; integrate gender concerns and perspectives in policies and programs for sustainable development; and strengthen or establish mechanisms at the national, regional and international levels to assess the impacts of development and environmental policies on women (United Nations 2008). In decisions explicitly on gender, the original decisions (36.CP.7 and 23/CP.18) are framed solely around enhancing women's participation and the gender balance in UNFCCC decision-making. However, decision 23/CP.18 called for a workshop at COP19 to address gender-sensitive climate policy, as well as mandate a standing agenda item on gender at the COP. The current conclusion under this agenda item (FCCC/SBI/2013/L16) includes proposals for workshops to further clarify the intersection of gender in all areas of the negotiations, as well as to monitor the implementation of gender sensitive climate policies and actions. Enhancement of institutional capacity to mainstream gender issues in global and national climate change policies and operations through the development of gender policies, gender awareness, internal and external gender capacity and expertise should be prioritized.

Women and men feel the impacts of climate change and disasters differently. A study by Neumayer/Plümper (2007) states that women and children are 14 times more likely to die or be injured in a disaster than men. During the 1991 cyclone in Bangladesh, reports state that 90% of the 140,000 killed were women (Neumayer/Plümper 2007). In addition to these staggering fatality statistics, women are affected by climate change and disasters more than men in other ways. The first section of this chapter will explore these differences and the factors contributing to women's differentiated vulnerabilities. It presents the links between vulnerability and inequality, with an emphasis on rights. The chapter will then discuss the responses and solutions in Bangladesh to these differentiated vulnerabilities, through government policy and projects and programmatic interventions and research. The chapter will conclude with a recommendation to move away from vulnerability discourse and towards a dialogue on rights and inequality, and highlight some opportunities and challenges for Bangladesh in its pursuit to battle the adverse impacts of climate change and disasters into the future.

8.2 Understanding Vulnerabilities and Rights

Development partners in Bangladesh have recognized that poverty is a key determinant of vulnerability (Oxfam 2008) and, as such, efforts have been made to try and increase the financial capacities of those in poverty to reduce their vulnerability. In many cases in Bangladesh, women are the target population for such interventions. Poverty reduction measures that target poor female-headed households, increasing the number of women entrepreneurs, and women-focused livelihood projects have all contributed in different ways to increasing women's economic situations. Yet, issues around the control of new financial gains within the household beg the question, are intra-household dynamics inhibiting women's financial decisions and hence their abilities to adapt to climate change? Further, when disaster hits, erosion of these financial gains for women and men set the process of poverty alleviation back. This leads to the belief that poverty reduction alone cannot redistribute the impacts of climate change equally between women and men and cannot solely contribute to building resilience to climatic change.

The lack of control and ownership of resources by women point to a persistent discrimination that has been recognized as causing increased vulnerability of women over men. Decision-making over resource use and the buying and selling of land pre-, during and post-disaster or slow onset climatic change, often rests with men. While women are often charged with the safeguarding of small-scale household goods, it is mostly men who make decisions around productive assets, such as when to sell what, to financially cope in times of environmental stress (International Food Policy Research Institute 2014). The ability of women to cope and bounce back is hampered compared to men, due to this discrimination. As a result, Sultana (2014) argues that issues of land rights and inheritance is essential in the discussion and debate on reducing this specific vulnerability for women. This is poignant in societies such as those in Bangladesh, where Islamic inheritance rights continue to discriminate against women, and religious norms result in sons receiving the majority if not all inheritance.

Individual nutritional status and access to clean, safe drinking water, which reflect intra-household food distribution, are other factors of discrimination that contribute to women's differential vulnerability to climate change in Bangladesh. Despite the basic human right to food, in Bangladesh women and often children tend to eat less in the household, compared to the men. Men eat first and in surplus, and women do not get their daily calorie requirement (World Food Programme 2012). In relation to water consumption:

…lack of safe water will affect the health and well-being of all members of a household, exacerbating household vulnerabilities and poverty. This strains the reproductive and care-giving roles of women. These are some of the ways that climate-induced ecological change affects men and women differently. (Sultana 2014: 375)

Furthermore, during times of disasters, it is often women's different needs and requirements including safe latrine facilities, private bathing facilities with safe and clean water and sufficient sanitary supplies that relief efforts fail to meet (Arora-Jonsson 2011). These issues of food, health and sanitarian also contribute to the increased vulnerability of women.

Violence and social isolation are also contributing factors to women's increased vulnerability compared to men in the context of climate change and disasters. Violence is a reality of most Bangladeshi women's lives: 87% of ever-married women have experienced some form of violence (Bangladesh Bureau of Statistics 2011). In Bangladesh, during disasters women and girls frequently suffer intimidation, violence and sexual harassment both moving to and in shelters. A study by Nasreen (2008) found that 71.6% of women respondents were subject to more violence during disasters. 'Many women and girls do not take refuge in shelters during disasters due to a lack of personal security' (Ahmed et al. 2015: 21). Despite the normalization of violence in Bangladesh, this violence is a fundamental human rights abuse against women that increases in frequency because of climate change.

Finally, women's lack of mobility in public spaces and consequent social isolation results in women not receiving early warnings, either at all, or in time. The right to information often goes unrealized for women in rural climate-affected regions. In some instances, women do receive the information, yet their unequal position in the household means they lack the decision-making power to determine what to do with this information they receive. These compounded factors lead to women's increased vulnerability to climate change and disasters.

The sum of these factors and disproportionate vulnerabilities can be attributed to the subordinate position of women in society, the lack of rights realization and the pervasive and persistent gender inequalities that exist in Bangladesh. Prevalent gender inequalities and power differences limit women's ability to respond and adapt to disasters and climate change impacts. It is the inequalities in the everyday, and not just in times of disaster, that create greater risk and reduces life chances for women and girls (UN Women 2015). Therefore, addressing gender inequality will lead to a reduction of vulnerability for women and create space to build resilience for both women and men to face the adverse impacts of climate change and disasters.

Acknowledging that the links between gender and climate change are complex and dynamic, attempts are underway to tackle women's increased vulnerability in Bangladesh and around the world. Yet, dialogue continues to focus on differentiated vulnerabilities of women and men, not the issue of rights (Intergovernmental Panel on Climate Change 2014). While rights abuses exist, as does gender discrimination, efforts in Bangladesh from the national and community levels are being made to help both women and men to respond and adapt to the changing climate. The next sections of this chapter will present these efforts and explore some of the ongoing research and project findings that are working towards building breadth and depth to the understanding of the differentiated impacts of climate change on women and men. It is through this understanding that more concrete solutions can be found and begin to change the situation for women on the ground.

8.3 National Efforts to Address Gender and Climate Change

The climate change and disaster management policy space is vast in Bangladesh. Numerous government ministries and cross-ministerial taskforces are charged with managing the current and predicted impacts of the changing climate. While the majority of ministries have climate change focal points and many have dedicated teams to integrate issues of disaster planning and management into their projects and policies, the *Ministry of Environment and Forests* (MoEF) and the *Ministry of Disaster Management and Relief* (MoDMR) are the technical leads on their respective topics with the government. The two key instruments, the *Bangladesh Climate Change Strategy Action Plan* (BCCSAP) 2009, funded through the Government's Bangladesh Climate Change Trust Fund,[1] and the National Disaster Management Plan 2010–2015[2] sit within the ministries and are used to guide planning and programing.

Within the national framework for addressing climate change (the BCCSAP), the Government of Bangladesh has made efforts to recognize and begin to address the increased vulnerability of women in climate and disaster affected regions of the country. Gender mainstreaming currently exists, to some extent on paper, across this strategy document with some gender perspectives integrated into plan. This includes specific action in Strengthening of Gender Considerations in Climate Change Management, highlighting the importance of addressing the needs of the poor and vulnerable, including women and children, in all activities under the plan (Ministry of Environment and Forest 2013) (An analysis of the gender perspectives of such strategy documents is provided in Box 8.2). Despite being created in 2009, efforts to specifically address the needs of women have been slow; as a result, development partners joined with MoEF to focus on this issue. To operationalize and prioritize implementation of the specific gender issues of the BCCSAP, the MoEF, with technical support from the IUCN, produced the *Bangladesh Climate*

---

[1]Another funding source, the Bangladesh Climate Change Resilience Fund (BCCRF), is a financing mechanism for development partners. Established in 2010, to date it has channeled US $188 million in grant funds (Bangladesh Climate Change Resilience Fund 2013). The fund is managed and implemented through the Government of Bangladesh. At this point, no systematic efforts or mechanisms exist to ensure these funds contribute to the reduction of gendered vulnerabilities and the differentiated impacts of climate change on women.

[2]The new National Disaster Management Plan for 2016-2020 was in final stages of preparation at the time of writing.

Change and Gender Action Plan (ccGAP) in November 2013. To date, this action plan has not been implemented and any specific gender related outcomes from projects operational under the BCCSAP remain unknown.

Box 8.2: Gender Impact Assessment of Bangladesh Climate Change Policies and Pathways for Mainstreaming by Mahin Al Nahian, GM Tarekul Islam and Sujit Kumar Bala.

The impacts of climate change are gender specific at different scales. The *Hyogo Framework for Action* (HFA) and the *United Nations Framework Convention on Climate Change* (UNFCCC) asked for gender inclusion in all types of climate change actions, but in reality, achievement is far from satisfactory. Bangladesh is currently leading its climate change negotiation and adaptation through the Bangladesh *Climate Change Strategy and Action Plan* (BCCSAP) and *National Adaptation Programme of Action* (NAPA). We tried to assess these policies using the '*Gender Impact Assessment* (GIA) Tool' to produce recommendations for gender mainstreaming in future revisions. GIA, a core tool for gender mainstreaming, helps to estimate the effects of a policy or activity in terms of gender equality; it can be highly effective in identifying gaps and suggesting recommendations for mainstreaming (Sauer 2010). The policies were assessed in terms of their positive, neutral and negative impacts in relation to gender and climate change. The GIA of the NAPA (Ministry of Environment and Forest 2009b) showed that the policy document acknowledged the need of gender inclusion as an important aspect of adaptation. Gender perspective was suggested as one of the major criteria for prioritization of adaptation needs and activities, which is encouraging. The policy also sought to address some key practical gender needs. However, the involvement of female experts in preparing the NAPA was minimal and the gender dimension of vulnerability was not explored. It is also questionable how the gender perspective could be used without establishing the linkage between gender and climate change. There are no guidelines to use the gender equality approach and no activities were identified for climate refugees. The GIA of the BCCSAP indicated that the policy document encouraged the inclusion of a gender lens in designing and implementing climate change adaptation activities (Ministry of Environment and Forest 2009a). It kept space for the development of approaches for gender inclusion in future revisions which is very important. Though the policy document acknowledged that women would be the worst sufferers due to climate change, it did not elaborate on how climate change would impact women – especially their roles and associated vulnerabilities. The policy failed to provide any action plan to address gender needs and there was no pathway suggested on how women will be involved in adaptation activities. Based on the study key recommendations for gender mainstreaming were suggested: (a) formulation of gender policy with budgeting, (b) a core group for mainstreaming, (c) a gender disaggregated baseline and indicator to assess present situation and future progress (d) 'gender-just' governance, based on gender equality both horizontally and vertically, keeping youth-age community central in adaptation activities, and (e) 'engendering' of adaptation strategies before implementation, etc. The study also suggested a 'Bottom up – top support' institutional framework for climate change adaptation and mitigation keeping community at the centre of intervention. Gender should be integral in both climate change adaptation and mitigation (Kapoor 2011) – this study is expected to encourage pathway for gender mainstreaming in Bangladesh's future climate change adaptation and mitigation policies and activities.

Turning to disaster management, while there is no independent section on gender and disaster management in the NDMP strategy document, women are identified as 'a distinct target group and agent in disaster forecasting, preparedness and management' (Ministry of Women and Children's Affairs 2015: 50). A key achievement under this strategy is the reform that has taken place through the revised Standing Order on Disasters (2010). This ensures that women have an increased role in preparedness activities and disaster management through assigned seats for women on district, upazila and union disaster management committees; Cyclone Shelter Management committees also mandate women's participation. A further discussion on the success of this reform comes later in this chapter.

Despite the identification of women's needs and vulnerabilities in national frameworks for climate change and disaster management, operationalization and the implementation by government of gender sensitive or gender-responsive policies and programs remain limited. While there is no specified role outlined for the *Ministry of Women and Children's Affairs* (MoWCA) in the BCCSAP, the ccGAP has developed specific interventions to be lead and technically supported by MoWCA, which include issues of social protection, increased livelihoods for women affected by climate change and gender-responsive policies to name a few. Yet, MoWCA remains a relatively weak ministry, and there is a significant need to build technical capacity of this ministry to implement activities and recommendations of the ccGAP, as well as lead the process of gender mainstreaming and effective gender-responsive policy reform in climate change and disaster management.

Finally, a recent paper commission by the *General Economics Division* (GED) of the Ministry of Planning indicates more positive steps to understanding and addressing the gender dimensions of climate change, including recommendations to implement the ccGAP within the 7th 5 Year Plan of the Government of Bangladesh (Ahmed et al. 2015). The GED, who leads the process of developing the country's 5 Year Plans, is in the final stages of producing the 7th Plan. The thematic background papers, both the paper on Climate Change and Disaster Management and Gender Equality and the paper on Women's Empowerment speak to women and men's differentiated vulnerabilities and the need for the State to invest to reduce women's specific vulnerability to climate change. If these key issues make it into the final plan, resources and targets will be allocated to such activities and hopefully government efforts will begin to reduce the specific vulnerabilities of women affected by climate change.

8.4 Decentralized Policy at the Community Level

Bangladesh has a reputation among lesser developed countries as a leader in community level disaster risk reduction; it was one of the first countries to establish community level disaster risk management processes (Ikeda 2009). Yet, when community-based procedures come from national agencies and are translated in local contexts, we can see gender concerns can be lost, diluted and often ignored in the operationalization of the policy (Ikeda 2009). Women's differentiated vulnerability is recognized in community-based initiatives, but there is a lack of effective response to address this vulnerability; 'existing policies do not consider gender-specific operational activities (Shabib and Khan 2014: 332).

Women are often marginalized or silenced in community projects. As a result, the gendered implications of climate change can be further exacerbated by uncritical conceptualization and implementation of adaptation programs that come in the name of community. The need to create space for different voices and recognition of a multiplicity of opinions and concerns could not be more urgent, so that men and women can all benefit from climate adaptation programs (Sultana 2014: 378).

Examples from efforts in Bangladesh unfortunately show similar results at the community level. Reform of the Standing Order on Disasters 2010 is a prime example. A minimum number of female representatives and a specific role for a female member responsible for vulnerable groups was introduced to increase participation of women in local and regional (Upazila) level disaster management committees. Yet, the results of such reform in addressing women's increased vulnerabilities remains to be seen. In fact, a 2015 study by UN Women and the Bangladesh Centre for Advanced Studies (BCAS) explores this new role of women in the local level disaster management committees; the study aimed to understand how and why issues of gender are seemingly still not being addressed in the process of implementing community disaster risk management. The study concluded that stereotypical gender roles for women, including caring for children, collection of food, fuel and water were all contributing factors inhibiting women's full and active participation in local level disaster management committees (UN Women and BCAS 2015).

Another study (Ikeda 2009) investigates community level actors and how they shape issues of gender. This study is unique in that it not only looks at local level government representatives and union members, but also religious leaders, local elite and the NGO staff who support communities. The study explores the translation of gender concerns from national project and policy levels, and identifies who shapes these issues at the local level. Not surprisingly, male community members in positions of power are most often the ones translating policy and governance into action at the community level. From both the UN Women/BCAS study and Ikeda's research, one can see a watering-down of the gender policies/issues and ideals from the higher national administrative layers to what happens on the ground. More research and analysis is required to better understand how gender issues are translated and operationalized in local level adaptation, disaster risk reduction policies and programs, and what interventions from government and non-government actors are required to achieve more concrete gender outcomes and results for women on the ground.

8.5   Civil Society and Development Efforts: Research and Practice

Civil society and development partners in Bangladesh have made significant efforts to reduce the impacts of climate change and disasters on communities. In particular, *community-based adaptation* (CBA) has been a very common strategy for implementing region-specific strategies to assist both women and men in their fight against cyclones, floods and droughts. However, too often the 'community' approach fails to recognize the discriminations against women. This next section presents research on CBA, as well as migration, and highlights a recent study paper stating how economic loss and financial stress from climate change is resulting in early or child marriage being used as an adaptation strategy.

Wong's (2009) study of a *United Nations Development Programme* (UNDP) home solar system project in the south-eastern part of Bangladesh draws out how community-based projects don't naturally benefit both sexes. This project established village level governance structures to oversee and implement the home solar systems in the Hindu community of Chokoria. Despite the commitment of the implementing civil society organization and UNDP to women's empowerment, and their requirement for specific roles and responsibilities to be delegated to women, the institutionalization of women's participation in a patriarchal society proved very difficult. Women and men were assigned particular jobs at the community level, many of which reinforced gender stereotypes but were also agreed to and went unchallenged by women and men. Women even justified their subordinate position and their suitability for lesser roles in committees. For example, all decision-making and external relations roles were assigned to men (Wong 2009). The project and Wong's subsequent research explored how attempts to change social norms and practices through community-based initiatives can be challenging and that institutional structures that enable gender inequality require more significant efforts for reform.

Another study, a collaborative research project by Oxfam Australia, Oxfam Great Britain and Monash University in conjunction with local partners in Bangladesh, has aimed to assess the gendered implications of climate change more broadly (Alston et al. 2014). The research revealed that changing agricultural patterns is resulting in out-migration of males from communities. This is altering intra-household dynamics (Alston et al. 2014), however not necessarily in favor of women. This issue of migration as a coping mechanism to climate change is beginning to be investigated by researchers in Bangladesh; however, projects and efforts specifically designed to minimize negative impacts of migration are yet to be seen.

Male migration is becoming quite common and, although reports state that environmental change is not the sole reason but one of a number of reasons for migration, migration *is* being viewed as a coping mechanism in response to environmental stress (Kartiki 2011) (see Box 8.3 for more examples of coping strategies used by women and men in Bangladesh). UN Women and the *Bangladesh Centre for Advanced Studies* (BCAS) completed a research project to understand the impacts of male migration on the women who remain behind (2015). The study looked at 10 districts across the country and found differing impacts of migration. In most cases, men tended to migrate to a city center or the capital, and the hardships and increased burden in women's lives only grew. In many cases, men didn't earn enough money to send home, or were not willing to, and the women were charged with existing household chores as well as finding paid work to provide basic needs for the family. Given the little available work (which led to the male out-migration initially) and gender pay discrimination, financial hardship was reported as common. As articulated in Fig. 8.1, migration patterns influence gender and gender consequently has influence over migration patterns.

From the UN Women and BCAS study, it was reported that community attitudes towards females with migrating husbands changed, and numerous women reported increases in harassment from other men and boys in the village (UN Women and BCAS 2015). However, some positive benefits of migration were reported, with women stating that when money was sent home, it was more than could have otherwise been made in the village. In some cases, an increase in women's decision-making abilities were reported and this has the potential to be a positive step towards changing household gender roles. However, the uncoordinated nature of migration, consequential rights abuses and negative impacts for women who remain behind warrant further attention.

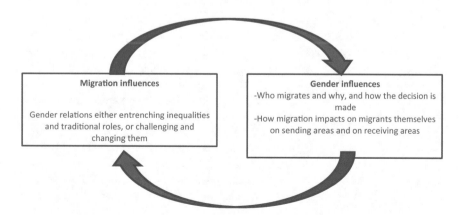

Fig. 8.1 Gender-Migration relationship. *Source* Institute of development studies (2005)

Box 8.3: Climate Change, Environmental Stress, Gender and Livelihood Resilience: Findings from Bangladesh by Sonja Ayeb-Karlsson, Kees Van Der Geest, Tanzinia Khanom, Istiakh Ahmed, Saleemul Huq, Sarder Alam Shafiqul, Koko Warner.

This case study is based on findings from the Gibika research to action project. The five-year effort aims to enhance livelihood resilience in seven project sites in Bangladesh. Throughout the research, data have been gathered for men and women separately to identify gender differences (Ayeb-Karlsson et al. 2016). Previous studies show that certain socio-cultural factors make women more vulnerable to climatic stress. For instance, it is sometimes considered unsafe or inappropriate for women to leave the house and evacuate to shelters during floods and cyclones for various reasons (Cannon 2002; Dankelman 2008; Sharmin and Islam 2013). Our fieldwork revealed similar trends in the coastal study sites.

The research findings show that men and women have different perceptions of climatic stress and how to address its impacts. Men tend to confront environmental stress by switching to alternative livelihood activities, while this is not as common among women. In many rural areas of Bangladesh, especially in the South, women's work is predominantly house-based. There tends to be disapproval towards women working outside of the house (Biswas et al. 2015; Sharmin and Islam 2013). The socio-cultural norm is that women are responsible for the house and men for the livelihood and economy. Socio-cultural constraints do not only affect women but also men (Ayeb-Karlsson et al. 2015). A man who is not capable of providing enough income, and whose wife has to work outside, receives less respect in the community. A good woman does not work outside the house, and a good man makes sure that his wife does not need to leave the house to work. Women in Bangladesh are not helpless victims, but work actively in reducing their vulnerability to impacts of climate change and extreme weather events. It is important to critically examine the underlying causes.

The research findings show that women search and find ways to adapt and increase their resilience and that their strategies tend to differ from those of men. In the central and North-eastern study-sites women were involved in alternative livelihood activities such as cutting mud, constructing roads, handicrafts and carrying bricks. Even though women in the South were more constrained in their movements and work possibilities, they were in some ways more prepared against disasters than men. As men were occupied with their work, more women than men attended disaster preparedness meetings and trainings. Women were also involved in other disaster risk reduction initiatives, such as sharing of livestock and ponds in the drought affected study sites. On top of that, according to other studies women are more involved in rehabilitation and post-disaster reconstruction work while men tend to migrate temporarily to cope with the shock (Biswas et al. 2015; Enarson 2000; Mallick 2014). To conclude, even though women in Bangladesh are more vulnerable to climatic stress than men, our research reveals a high level of resilience and determination in how they face the environmental threats.

A recent study by Alston et al. (2014) investigated the impacts of climate change on women and found that physical safety and school drop outs became an issue in parts of Bangladesh. The study reported an increase in violence against women and highlighted women's disproportionate suffering from food shortages and a decrease in the number of girls in education. An unexpected key finding was an increase in early marriage, where the dowry is viewed in society as a source of capital and incredible valuable in times of crises (Alston et al. 2014). In fact, the link between the changing climate and child marriage was recently picked up by Human Rights Watch. The 2015 Human Rights Watch Report titled 'Marry Before Your House is Swept Away' illustrates numerous examples on why marriage is seen as essentially an adaptation strategy to climate change and disasters. The continued loss of assets in floods and river erosion, the lack of financial stability to feed families and the relatively high price for young bride dowry are all contributing factors to marry off girls at a young age (Human Rights Watch 2015).

While there has been an increased focus on women in the climate change and disaster arenas (IIED 2015), including in Bangladesh, the realities of women's lives and the pervasive gender discriminations are often not addressed in climate change adaptation and disaster preparedness, or response and recovery initiatives. Gendered stereotypes underpin most of the climate change and disaster risk reduction initiatives and focus is often on the practical needs of women as wives and mothers (IIED 2015), rather than the rights of all equal citizens. Policy efforts described in this chapter are an example of this. Some interventions target women as vehicles for supporting households or strengthening community systems, as seen in Wong's (2009) study, but seldom do initiatives directly address gender inequalities at the societal, community, or household level. This is where the opportunities and challenges lie ahead for actors in Bangladesh.

8.6   Further Opportunities and Challenges

Climate change is predicted to continue and so will the adverse impacts on women and men. Unless the root cause of such gendered vulnerabilities and human rights abuses are addressed, these adverse impacts may only worsen the situation for women in Bangladesh. This final section details opportunities and challenges for stakeholders in their efforts to reduce the human impacts climate change.

The Government of Bangladesh has an opportunity now, through the implementation of the 7th 5 Year Plan, to actively address issues of human rights and gender equality as they relate to the changing climate in Bangladesh. At this point, little evidence exists that community-based examples and lessons learned, such as those explored in this chapter, are feeding into national level policy and decision-making. In addition, traditional conceptions of gender roles continue to dominate much of the discussion on the gendered dimensions of climate change. The lessons learned from development initiatives and research need to target policymakers and provide recommendations for action.

In international climate change and disasters discourse, women have been characterized as 'victims', 'most vulnerable' and in need of protection; rarely are women given the chance to articulate their rights, needs and responsibilities (Tschakert and Machado 2012). Women are often seen as beneficiaries and targeted to be the agents that assist in delivering interventions. This is true for the case of Bangladesh, where women are identified as vulnerable and in need of protection, as explored in this chapter through analysis of both national frameworks and in local level adaptation and disaster management activities. Yet, limited dialogue focuses on women as equal rights holders or equal citizens alongside men. Hence, the author argues there is a need to change how women are seen in climate change and disaster discourse, 'moving away from descriptive vulnerability assessments to diagnosing drivers of inequality, marginalization, and barriers to transformative change, and promoting agency and resilience through processes of engagement and collective learning' (Tschakert and Machado 2012: 278).

Efforts to address issues of gender equality in climate change are currently limited to a few projects. With the increasing interest and advancing scholarship which draws out key themes such as women's participation in community efforts, migration and child marriage, there is an opportunity for civil society organizations and development partners to take this knowledge and work directly with women and men. Moving from analysis of vulnerabilities to debates of rights requires a climate change discourse to focus on the social dimensions, rather than a technological or economic focus, and 'humanizing the complexity of the problem incorporating fairness and responsibilities, while also providing room for engagement' (Tschakert and Machado 2012: 280). The opportunity is now for actors to focus their efforts on the human rights and gender equality dimensions of climate change to improve the lives of women and men in Bangladesh.

References

Ahmed, A.U., Haq, S., Nasreen, M. and Hassan, A.W.R. (2015). *Sectoral inputs towards the formulation of Seventh Five Year Plan (2016–2021)* Climate Change and Disaster Management, Dhaka Bangladesh.

Alston, M., Whittenbury, K., Haynes, A. and Godden, N. (2014). Are climate challenges reinforcing child and forced marriage and dowry as adaptation strategies in the context of Bangladesh? *Women's Studies International Forum* (47), 137–144.

Ayeb-Karlsson, S., van der Geest K., Ahmed I., Huq S. and Warner, K. (2016). A people-centred perspective on climate change, environmental stress, and livelihood resilience in Bangladesh. *Sustainability Science*, 11(4), 1–16.

Ayeb-Karlsson, S., Tanner, T., van der Geest, K., Warner K. et al. (2015). *Livelihood resilience in a changing world – 6 global policy recommendations for a more sustainable future.* UNU-EHS Working Paper 22. UNU-EHS: Bonn.

Arora-Jonsson, S. (2011). Virtue and vulnerability: Discourses on women, gender and climate change. *Global Environmental Change*. 21(2), 744–751.

Bangladesh Bureau of Statistics (2011). Report on Violence Against Women Survey 2011, Bangladesh Bureau of Statistics, Ministry of Planning, Government of Bangladesh.

Bangladesh Climate Change Resilience Fund (2013); at: http://bccrf-bd.org/ (9 July 2015).

Biswas, A.A., Zaman, M., Sattar, M.A., Islam, M.S., Hossain, M.A. and Faisal, M. (2015). Assessment of Disaster Impact on the Health of Women and Children. *Journal of Health and Environmental Research* 1(3):19-28.

Cannon, T. (2002) Gender and climate hazards in Bangladesh. Climate Change and Gender Justice. *Gender & Development* 10(2):45-50.

Dankelman, I.E.M. (2008). *Gender, climate change and human security: Lessons from Bangladesh, Ghana and Senegal.* External Research Report. New York: WEDO.

Enarson, E.P. (2000). *Gender and natural disasters. InFocus Programme on Crisis Response and Reconstruction* Working Paper 1. Geneva: ILO.

Human Rights Watch (2015). Marry Before Your House is Swept Away; at: https://www.hrw.org/report/2015/06/09/marry-your-house-swept-away/child-marriage-bangladesh (10 August 2015).

Ikeda, K. (2009). How women's concerns are shaped in community-based disaster risk management in Bangladesh, *Contemporary South Asia*, 17(1), 65–78.

Institute of Development Studies (2005). *BRIDGE Gender and Development in Brief*, Issue 16. Brighton, UK.

Intergovernmental Panel on Climate Change (2007). *Fourth Assessment Report (FAR) Climate Change 2007.* Cambridge - New York: Cambridge University Press.

Intergovernmental Panel on Climate Change (2014). *Climate Change 2014: Mitigation of Climate Change. Contribution of Working Group II to the Fifth Assessment Report of the Intergovernmental Panel on Climate Change* [Edenhofer, O. R. Pichs-Madruga, Y. Sokona, E. Farahani, S. Kadner, K. Seyboth, A. Adler, I. Baum, S. Brunner, P. Eickemeier, B. Kriemann, J. Savolainen, S. Schlömer, C. von Stechow, T. Zwickel and J.C. Minx (eds.)]. Cambridge University Press: Cambridge - New York.

International Institute of Environment and Development (IIED) (2015). Building resilience to environmental change by transforming gender relations: at: http://pubs.iied.org/17237IIED (30 April 2015).

International Food Policy Research Institute (2014). Enhancing Women's Assets to Manage Risk under Climate Change: Potential for Group-Based Approaches http://www.pim.cgiar.org/2014/11/25/enhancing-womens-assets-to-manage-risk-under-climate-change-potential-for-group-based-approaches/ (10 June 2015).

Kapoor, A. (2011). *Engendering the Climate for Change: Policies and practices for gender-just adaptation*, Alternative Futures and Heinrich BÖll Foundation (HBF).

Kartiki, K. (2011). Climate change and migration: a case study from rural Bangladesh, *Gender & Development* 19(1), 23–38.

Mallick, B. (2014). Cyclone-induced migration in southwest coastal Bangladesh. *ASIEN* 130: S60-81.

Maplecroft (2015). Climate Change and Environmental Risk Atlas (CCERA); at: https://maplecroft.com/portfolio/new-analysis/2014/10/29/climate-change-and-lack-food-security-multiply-risks-conflict-and-civil-unrest-32-countries-maplecroft/ (9 July 2015).

Ministry of Environment and Forests (2009a). *Bangladesh Climate Change Strategy and Action Plan 2009*, Ministry of Environment and Forests (MoEF), Government of Bangladesh.

Ministry of Environment and Forests (2009b). *National Adaptation Programme of Action* (NAPA). Ministry of Environment and Forests (MoEF). Government of Bangladesh.

Ministry of Women and Children Affairs (2015). *Bangladesh Report: The Implementation of the Beijing Declaration and Platform for Action (1995) and the outcomes of the Twenty-third Special Session of the General Assembly (2000)*, Dhaka, Bangladesh.

Nasreen, M. (2008). *Violence against Women in Floods and Post Flood Situations in Bangladesh*, ActionAid, Dhaka.

Neumayer, E. and Plümper, T. (2007). The gendered nature of natural disasters: the impact of catastrophic events on the gender gap in life expectancy, 1981–2002. *Annals of the Association of American Geographers*, 97(3), 551–566.

Oxfam (2008). Rethinking Disasters: *Why death and destruction is not nature's fault but human failure*. Oxfam International, New Delhi, India.

Rahman, M.K., Paul, B.K., Curtis, A. and Schmidlin, T.W. (2015). Linking Coastal Disasters and Migration: A Case Study of Kutubdia Island, Bangladesh. *The Professional Geographer*, 67 (2), 218–228.

Sauer, M.A.A. (2010). *Gender Impact Assessment integrated in Social Impact Assessment – the European experiment in sub-ordination*. Paper presented at EASY–ECO Conference, Université Libre de Bruxelles, Brussels, 17–19 November 2010.

Shabib, D. and Khan, S. (2014). Gender-sensitive adaptation policy-making in Bangladesh: status and ways forward for improved mainstreaming, *Climate and Development*, 6(4), 329–335.

Sharmin, Z. and Islam, M.S. (2013). *Consequences of climate change and gender vulnerability: Bangladesh perspective*. Bangladesh Development Research Working Paper Series. BDRC: Dhaka;at: http://www.bangladeshstudies.org/files/WPS_no16.pdf (2 March 2017).

Sultana, F. (2014). Gendering Climate Change: Geographical Insights. *The Professional Geographer*, 66(3), 372–381.

Tschakert, P. and Machado, M. (2012), Gender Justice and Rights in Climate Change Adaptation: Opportunities and Pitfalls. *Ethics and Social Welfare*, 6(3), 275–289.

United Nations (2008). 52nd Session of the Commission on the Status of Women Interactive Expert Panel. *Emerging Issues, Trends and New Approaches to Issues Affecting the Situation of Women or Equality Between Women and Men: Gender Perspectives on Climate Change*. February 2008 Issues paper.

United Nations Development Program (2009). Resource Guide on Gender and Climate Change; at: http://www.un.org/womenwatch/downloads/Resource_Guide_English_FINAL.pdf.

United Nations Development Program (2014). *Human Development Report 2014: Sustaining human progress, reducing vulnerabilities and building resilience*. UNDP, New York.

United Nations Framework Convention on Climate Change (2005). *Global Warning: Women Matter. UNFCCC COP Women's Statement*; at: http://www.genanet.de/fileadmin/user_upload/dokumente/Gender-Klima-Energie/Gender_CC_COP11_statement_short_7dec.pdf.

United Nations Framework Convention on Climate Change (2014). Bangladesh experiences with the NAPA process; at: http://unfccc.int/adaptation/knowledge_resources/ldc_portal/bpll/items/6497.php (7 July 2015).

United Nations Framework Convention on Climate Change (2014). *Existing Mandates and Entry Points for Gender Equality Technical Guide for COP20, Lima, Peru*; at: http://www.wedo.org/wp-content/uploads/GE-Publication-ENG-Interactive.pdf.

United Nations Office for Disaster Risk Reduction (2015). *Sendai Framework for Disaster Risk Reduction 2015–2030*; at: http://www.unisdr.org/we/inform/publications/43291 (9 July 2015).

UN Women (2015). *Gender, Climate Change and Disaster Risk Reduction and Recovery Strategy-Asia Pacific*. Suva, Fiji.

UN Women and the Bangladesh Centre for Advanced Studies (2015). *Climate Change and Migration in Bangladesh: A Gender Perspective*, UN Women, Dhaka.

Wong, S. (2009). Climate change and sustainable technology: re-linking poverty, gender, and governance, *Gender & Development* 17(1), 95–108.

World Food Program (2012). *Nutrition strategy*. Dhaka, Bangladesh.

Chapter 9
Pathways of Climate-Resilient Health Systems in Bangladesh

**Muhammad Abdur Rahaman, Mohammad Mahbubur Rahman
and Syed Hafizur Rahman**

Abstract Climate change is a complex phenomenon that will have a range of both anticipated and unexpected direct and indirect effects. The IPCC Fifth Assessment Report (AR5) affirms that recent decades have seen warming air and ocean temperatures, changing rainfall patterns, variations in the frequency and intensity of several extreme events including droughts, floods and storms and rising sea levels. The changing climate will adversely affect the health of human populations. These include primary or direct effects (e.g. increased deaths due to extreme weather events like cyclones); secondary or indirect effects (e.g. increased health problems due to disease vectors, such as malaria-carrying mosquitos and contaminated food and water); and tertiary or long-term effects (e.g. distractions for health and social services). This chapter provides an introduction to the relationship between climate change and human health, using the country-specific example of Bangladesh. Bangladesh is a low-lying country in which extreme climatic events are a common phenomenon. With the objective of providing an overview of the likely health impacts caused by climate change, the chapter examines the relationship between three distinct climatic events – flooding, salinity intrusion and drought – in relation to human health. In Bangladesh, issues such as poor water quality, unhygienic environmental conditions and poor sanitation, exacerbate the impact of climate-sensitive diseases (diseases of which transmission is linked to climatic and weather conditions). This chapter provides a foundation for studying the relationship between the climatic characteristics of the study area, climate-sensitive diseases and other anthropogenic phenomena. It demonstrates the pathways of climate-resilient health systems in Bangladesh.

Keywords Climate change · Climate extremes · Climate-resilient
Health system · Bangladesh

Muhammad Abdur Rahaman, Climate Change Adaptation, Mitigation, Experiment & Training (CAMET) Park, Noakhali, Bangladesh, Corresponding Author, e-mail: rana.bries@gmail.com.

Mohammad Mahbubur Rahman, Network on Climate Change, Bangladesh (NCC,B) Trust, Dhaka.

Syed Hafizur Rahman, Department of Environmental Sciences, Jahangirnagar University, Savar, Dhaka, Bangladesh.

© Springer Nature Switzerland AG 2019
S. Huq et al. (eds.), *Confronting Climate Change in Bangladesh*,
The Anthropocene: Politik—Economics—Society—Science 28,
https://doi.org/10.1007/978-3-030-05237-9_9

9.1 Introduction

There is increasing global concern about the impacts of climate change on human health. Climate change related hazards, such as drought, flooding, waterlogging, tidal inundation, cyclones, storm surges, erratic rainfall and rising temperatures, are common events in Bangladesh. They have direct and indirect adverse impacts on water resources, agriculture, livelihoods, ecosystems and human health.

9.1.1 Climate Change in Bangladesh

Bangladesh is situated in the world's largest and most populous Ganges–Brahmaputra delta system (Hossain et al. 2010). The country is considered the most vulnerable to tropical cyclones, the third most vulnerable to sea level rise in terms of the number of people affected, and the sixth most vulnerable to floods in the world (Francis and Maguire 2016). Scientists have proved that the erratic nature of rainfall and temperature is gradually increasing in Bangladesh. Precipitation is becoming less predictable and the monsoon is now characterised by higher amounts of rainfall within shorter periods of time (Islam et al. 2014). Temperatures are becoming more extreme, with regional variations and an overall annual rise. Tropical cyclones are also expected to increase in intensity.

As a low-lying country, Bangladesh is highly vulnerable to these climatic changes and extremities. Climate and weather patterns and behaviour play a significant role in freshwater availability, agriculture, economic growth and performance, and livelihoods (NAPA 2009). The most damaging effects come from flooding, drought and heat stress (World Bank 2013). The adverse effects of these on agricultural productivity and availability of freshwater are already evident in many areas of Bangladesh. Crop productivity is much reduced by drought, and perennial trees and livestock are damaged and lost as a result of floods every year. Furthermore, the increase in cyclone activity, combined with sea-level rise (SLR), will increase the depth and risk of inundation from floods and storm surges and reduce the area of cultivable land (particularly in low-lying deltaic regions like Bangladesh) (World Bank 2013).

9.1.2 Climate Change and Health

The relationship between climate change and human health is multidimensional. The health and wellbeing of human populations are sensitive to shifts in weather patterns and other aspects of climate change (IPCC 2014). There are three basic pathways by which climate change affects health, which can be classified as primary or direct impacts, secondary or indirect impacts, and tertiary or long-term implications respectively (see Butler 2014). Significant effects are those that affect people directly, such as heat stress and trauma from exposure to serious weather events (Butler/Harley 2010). Such effects are increasingly recognised and understood (Bowles et al. 2014). Secondary effects act less directly, through ecosystem change and other synergies, including shifts in the occurrence and predominance of diseases and vectors (Butler/Harley 2010). However, tertiary health effects, which are mediated through impacts on social, political, and economic systems, will likely have the greatest long-term impact on human health. This possibility remains relatively unexplored and scarcely recognised (Butler/Harley 2010).

The causal connections between climate change and human health are complex, as they are usually indirect, context-specific and dependent on specific climatic phenomena (Fig. 9.1). People's health depends on the ecosystem goods and services (such as availability of freshwater, food, wood, etc.) that are essential for sound human health and productivity (Corvalán et al. 2005). Notable direct human health impacts can occur if ecosystem services are no longer able to meet community needs. At the same time, changes in ecosystem services affect livelihoods, income and local migration and, at times, may even cause political conflict. The resulting impacts on economic and physical security, freedom, choice and social relations have wide-ranging effects on wellbeing and health, and the availability of ecosystem services (Fig. 9.1).

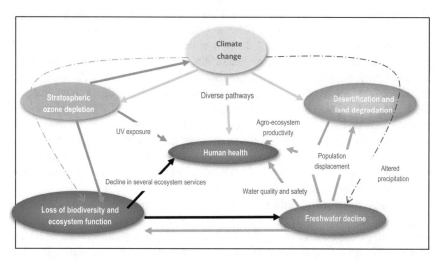

Fig. 9.1 Climate change threats to ecosystems and human health. *Source* Adapted from WHO (2016)

9.1.3   Context of Climate Change and Health in Bangladesh

In Bangladesh, climate change is having clear and significant mental and physio-logical impacts on human health (Rahman 2008; UNICEF 2011). The most direct effects of climate change on human health occur through extreme events. For example, in 2004, floods in Bangladesh caused 800 deaths and, in 2007, a sig-nificant cyclone (Sidr) affected more than 8.5 million people, causing more than 3,500 deaths (Shovo et al. 2013). Such extreme events result in both mental and physical health impacts. Slow onset climatic events also result in a multitude of health impacts. These are discussed in more detail throughout this chapter.

Climatic extremes such as heat stress also affect slum-dwellers. Over the last few decades, climate-induced migration of people has become inevitable in Bangladesh (Akter 2009). Many displaced people, having lost their land, home, and livelihoods to different climatic hazards, move to nearby small cities or distant larger cities like Dhaka to find a place to live and a livelihood, only to end up in a slum. Living conditions in these slums go from bad to worse with the rapid and uncontrolled increase of their populations (Saha 2012). Overall health and environmental con-ditions in urban slums are indigent, primarily due to overcrowding and lack of access to essential services, such as safe water and sanitation (Streatfield and Karar 2008). These poor conditions are exacerbated by climatic events such as frequent flooding and inundation, and waterlogging, which result in outbreaks of water-borne (diarrhoea, cholera, dysentery, jaundice, skin diseases) and vector-borne (dengue, malaria) diseases and other health problems (cold/cough, fever, pneumonia, pox, asthma, headaches) (Rashid et al. 2013).

But even within communities there are certain marginalised socioeconomic groups. For example, as a result of gender division, women are more vulnerable to climatic events than men. Other marginalised vulnerable groups include physically or mentally challenged persons, ethnic minorities (i.e., the tribal group of Chittagong hill tracts, Cox's Bazar and Mymensing) (CCC 2009b). Though some coping strategies for these vulnerable group have been discussed, concern for minorities was absent in Bangladesh's established plan for coping with climate change. Therefore, the state should prioritise different vulnerable social groups including women, youth and ethnic minority groups in planning coping mecha-nisms and adaptation strategies.

It should be noted at this stage that data on climate change and health in Bangladesh is limited. The Bangladesh Medical Research Council (BMRC), International Centre for Diarrhoeal Disease Research, Bangladesh (ICDDR,B), and Essential National Health Research (ENHR) carry out biomedical and operational research in Bangladesh. ICDDR,B has recently started a specialised unit called the Center for Population, Urbanization and Climate Change (CPUCC), with an aim to conduct research on human health and climate change. Nevertheless, there remains a lack of access to sufficient data, quality and availability of studies in this field, which must be addressed. This chapter reviews some of the data that is available to provide an overview of climate change and health in Bangladesh.

The chapter will focus on some of the indirect physiological impacts of climate change on human health in Bangladesh. It will first examine the relationship between climate hazards and climate-sensitive health afflictions and then look at the response processes that communities in Bangladesh have adopted to combat climatic stressors on human health.

9.2   Climate Related Specific Health Impacts in Bangladesh

9.2.1   Climate-Sensitive Diseases and Infections

One of the most poignant challenges of climate change in Bangladesh is its impact on climate-sensitive diseases and illnesses. These include malaria, dengue, typhoid and cholera, alongside skin diseases and newly identified rodent-borne diseases. Flooding can lead to increased rates of typhoid and cholera, while drought increases the risk of malaria and skin disease. Incidences of malaria have increased dramatically over the last 30 years (Alam 2011) (see Fig. 9.2).

Malaria is a mosquito-borne infectious disease induced by parasitic protozoans of the genus *Plasmodium* (*vivax, malariae, oval, knowlesi*, and *falciparum*) and is transferred by female mosquito vectors of the *Anopheles* species (Cox 2010). The spread of the disease is impacted by climate determinants and the local potential to manage it (Caminade et al. 2014). The impact of temperature on malaria prevalence appears to be nonlinear and is vector-specific (Alonso et al. 2011). Enhanced variation in temperature, when it is close to the uppermost boundary for vectors and pathogens, leads to decreased transmission of the disease, while increasing variations of daily mean temperature near the minimum limit for vectors and pathogens, increases transmission (Paaijmans et al. 2010). The nonlinear response to temperature means that even moderate warming may drive substantial increases in malaria transmission, if conditions are suitable (Alonso et al. 2011; IPCC 2014).

The average annual incidence of malaria in Bangladesh doubled from the period 1981–1990 to the period 1991–2000. It increased by another 65 percent between 2001–2010 and is now a major public health problem, with 13.25 million people

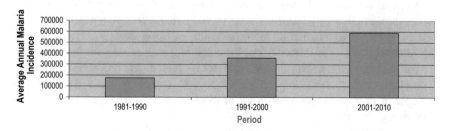

Fig. 9.2 Malaria trends in Bangladesh from 1981 to 2010. *Sources* DG-Health data (1997–2010)

across 13 districts in Bangladesh at risk of the disease (M0HFW 2015). Among these 13 districts, five districts (Rangamati, Khagrachari, Bandarban, Chittagong and Cox's Bazar) are at high risk, four districts (Mymensingh, Netrakona, Sherpur, and Kurigram) are at moderate risk and another four districts (Sylhet, Hobigonj, Sunamgonj, and Moulvibazar) are at low risk of malaria. The geophysical locations, climate, and other favourable conditions for the vector species transmission make these areas vulnerable (MoHFW 2015).

It is evident that variations in seasonal temperature and rainfall patterns significantly influence the mortality rate in Bangladesh. Each milimeter (mm) rise in average weekly precipitation up to 14 mm was associated with decreased mortality risks, while rainfall above 14 mm or below 3 mm was associated with increased mortality risk (Alam et al. 2012). Furthermore, the weekly mean temperature had the strongest relationship with weekly mortality rates. Below 23 °C, the relative mortality risk rose by 2.3 percent with each 1 °C reduction in temperature, and between 23 and 29.6 °C, the relative risk increased by 2.4 percent with a 1 °C decrease, particularly among females and adults (Alam et al. 2012).

There is evidence of an association between climatic patterns, such as the El Niño Southern Oscillation and the Indian Ocean Dipole, and dengue prevalence in Bangladesh (Banu et al. 2015). The connection between dengue occurrence and local climate variables such as temperature and precipitation are significant and apparent in the context of Bangladesh (Banu et al. 2015). Trends in dengue prevalence indicate peaks during the summer months. Similar trends are found in skin disease, while jaundice is common in flood prone areas due to the negative impact of flooding on safe water supplies.

The Nipah virus infection, a recently identified infectious disease that is sensitive to weather and climate variability, affects both animals and humans. Though the associations to climate change are indirect, the case of the Nipah virus is a clear indicator of the potential for increased infectious disease outbreaks as future climate change and human activities like large-scale deforestation modify animal habitats as well (Bennett and McMichael 2010). An outbreak of Nipah virus was reported in Bangladesh in 2013, with 24 cases and 21 deaths across 14 districts including Gaibandha, Natore, Rajshahi, Naogaon, Rajbari, Pabna, Jhenaidah, Mymensingh, Nilphamari, Chittagong, Kurigram, Kustia, Magura, and Manikganj. The infection spread to humans through date palm sap that had been contaminated through infected fruit bats (WHO 2013).

A causal connection between climate change and these diseases is difficult to verify, though climatic parameters such as temperature and precipitation are considered as key determinants of the distribution of many disease-carrying vectors (McMichael et. al. 1996). In addition, the conditions associated with the impacts of climate change on water supply, sanitation, and food production, generate favourable environments for the incidence and spread of such diseases. For example, a decline in the availability of clean water results in a greater risk of water-borne diseases.

In Bangladesh, the mortality impacts of water-borne diseases are likely to increase as climate change reduces the availability of safe drinking water (Reid and

Sims 2007). As climate change affects water quality, its impacts on clean water and sanitation facilities, along with poor hygiene, relate to various illnesses such as diarrhoea, cholera and dysentery (NIPORT 2015). For example, the occurrence of diarrhoea increased from 1995 to 2010. The highest incidences of diarrhoea in this period were seen in 1998, 2004, 2007 and 2010, which correspond to major flood years in Bangladesh (CCC 2009a; DG-Health 2010). It is important to note that the impacts of climate change on water resources vary depending on the climate-vulnerable area in Bangladesh. The effects of climate change on water resources in different climate-vulnerable areas of Bangladesh and their potential health impacts are outlined in Table 9.1.

Table 9.1 Typology of climate-vulnerable areas and potential health impacts in Bangladesh

Climate-vulnerable areas	Locations	Potential impacts of climate change	Risk intensity	Diseases and health impacts
Drought-prone areas	Northern and northwestern districts of Bangladesh	Depletion of groundwater level and quality, water scarcity, dehydration and heat stress, etc	Moderate, severe to very severe risk	Respiratory diseases (asthma), cardiovascular diseases (heart attack, heart failure, heart disease), heat stroke, diarrhoea, malaria, dengue, skin diseases, malnutrition and infectious diseases (kala-azar, Nipah virus)
Flood-prone areas	Ganges, Brahmaputra, and Meghna floodplain areas of Bangladesh	Drowning and inundation, water logging, lack of safe water, poor sanitation and hygiene, etc	Severe, very severe, and high-risk	Water-borne diseases (diarrhoea, cholera, dysentery, typhoid, jaundice, skin diseases), vector-borne diseases (dengue, malaria, filaria), fever and mental disorders

(continued)

Table 9.1 (continued)

Climate-vulnerable areas	Locations	Potential impacts of climate change	Risk intensity	Diseases and health impacts
Salinity-prone areas	Southern and southwestern coastal zone of Bangladesh	Scarcity of freshwater, high salinity in both surface and groundwater sources, etc	Very severe to high-risk	Diarrhoea, skin diseases, malnutrition, pneumonia, jaundice, diabetes, hypertension and reproductive health diseases

Source The authors

Climatic impacts on water resources vary in different locations. In drought-prone areas, groundwater quality depletes and water scarcity increases. Flood-prone areas become submerged during torrential rains and overflowing rivers. Flood water contaminates tube-wells, which means that they cannot supply drinking water at these times. Those living in flood-prone areas thus suffer from different climate-sensitive illnesses related to drinking water scarcity (quality and quantity). In coastal regions, salt water intrusion resulting from SLR affects groundwater and surface water, causing health burdens for coastal populations.

9.2.2 Climate Change and Malnutrition

Fish and livestock are the principal sources of protein and nutrition for both rural and urban people in Bangladesh. Without them, people suffer from malnutrition and poor health. However, in Bangladesh, agricultural, forestry and fishery systems are increasingly being affected by fluctuating SLR, cyclones, flooding and drought. During the 1998 floods, 69 percent of Aus rice production, 82 percent of deepwater Aman and 91 percent of transplanted Aman were lost, leaving the whole country food insecure (Dewan 2015). Each year, drought in Bangladesh affects approximately 2.32 million hectares of cropland during the *Kharif* (summer monsoon) and 1.2 million hectares of such land during the *Rabi* (winter) seasons (Sugden et al. 2014). Frequent cyclonic events, inland and coastal flooding, low water flow, droughts, salinity intrusion and changes in morphological processes, threaten existing aquatic ecosystems, fisheries, and livestock, thereby impacting fish and livestock production. Climate change and its impacts can result in the outbreak of new diseases and pests that will further affect these subsectors (IPCC 2014). Figure 9.3 illustrates the pathways through which malnutrition or undernourishment develop in the human body.

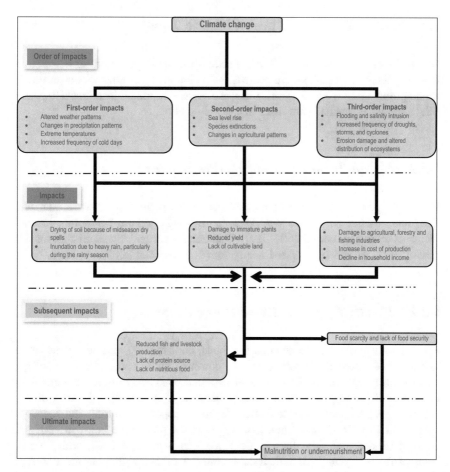

Fig. 9.3 Climate change impacts on agricultural systems in Bangladesh and pathways to malnutrition. *Source* The authors

Processing, distribution, acquisition, preparation and consumption activities are equally important for food security and are also impacted by changes in climate. For example, increased frequency and intensity of severe weather increase damage to transport and distribution infrastructure, with consequent disruption to food supply chains (FAO 2008). Ultimately these climatic impacts are reducing people's access to food and the specified range of nutrients, thereby affecting malnutrition levels in Bangladesh.

9.3   Health Trends in Three Climatic Zones

Outbreaks of dangerous, water-borne, vector-borne and other types of disease, are most prevalent in climate sensitive areas, including drought-prone, flood-prone and saline-prone areas (Fig. 9.4). The disease trends in three locations (Rajshahi, Sirajganj and Patuakhali) are discussed in this section.

9.3.1   Health Trends in Drought-Prone Areas

In drought-prone areas, health ailments such as diarrhoea, skin disease, malnutrition, and asthma are increasing (Dey et al. 2012). Figure 9.5 points to such increases during the period 2004–2010. This trend is particularly significant for skin disease, rates of which increased nearly four-fold between 2009 and 2010.

9.3.2   Health Trends in Flood-Prone Areas

The occurrence of diseases such as diarrhoea, cholera, skin disease, malnutrition, asthma, cold/cough and fever appear to correlate with events like flooding and water logging (Rahaman 2013). Skin disease and malnutrition trends are gradually increasing in flood-prone areas, mainly because of a shortage of safe drinking water, proper hygiene, and sanitation facilities, especially during the monsoon season. As Fig. 9.6 shows, when looking at disease prevalence between 2001 and 2010, diarrhoea and skin disease incidence was highest in 2007, which was a significant flood year in Bangladesh. Figure 9.6 further highlights the trends of some these health issues in the flood-prone area of Sirajganj during the period 2001–2010.

9.3.3   Health Trends in Salinity-Prone Areas

Diarrhoea, skin disease and malnutrition rates are increasing in the saline-prone areas of Bangladesh. This is because of the frequency of various natural disasters, including cyclones, storm surges, coastal flooding, sea level rise and salinity intrusion, which increases the burden of water-borne and vector-borne diseases. By looking at trends from 2001 to 2010, Figs. 9.8, 9.9 and 9.10 show that the prevalence of diarrhoea, malnutrition and skin disease is rising in Patuakhali, a saline prone coastal district of Bangladesh.

In the case of diarrhoea, the incidence rate was found to be higher during the monsoon and pre-monsoon periods, whereas it was comparatively lower during the

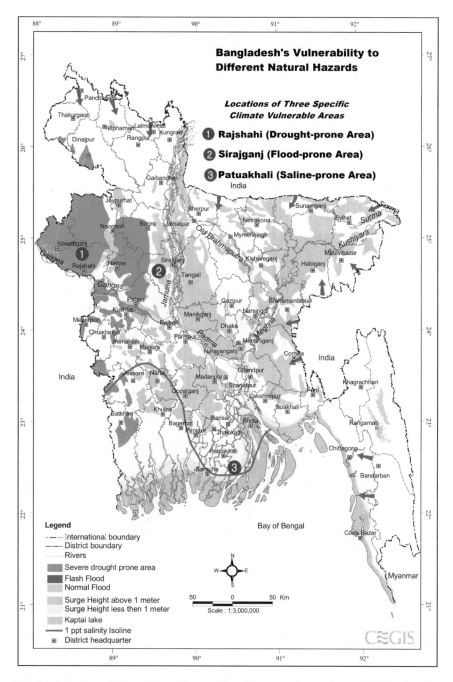

Fig. 9.4 Locations of three distinct climate-vulnerable areas: *Source* Authors OWN using data From CEGIS, Dhaka

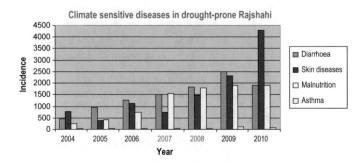

Fig. 9.5 Health trends in the drought-prone Rajshahi district from 2004 to 2010. *Source* Rahaman (2013)

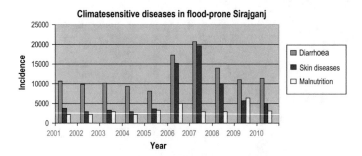

Fig. 9.6 Health trends in the flood-prone Sirajganj district from 2001 to 2010. *Source* Rahaman (2013)

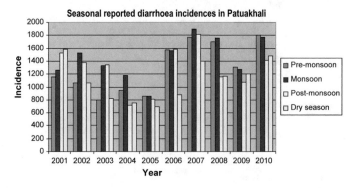

Fig. 9.7 Seasonal occurrence of diarrhoea in saline-prone Patuakhali district from 2001 to 2010. *Source* Rahaman (2013)

post-monsoon and dry seasons (Fig. 9.7). The overall disease occurrence rate increased at a significant rate (Fig. 9.7). Malnutrition rates were higher, especially during the monsoon (2004, 2007, 2008, 2009 and 2010) and post-monsoon (2001, 2002, 2004 and 2008) seasons (Fig. 9.8). As a result of flooding during this period,

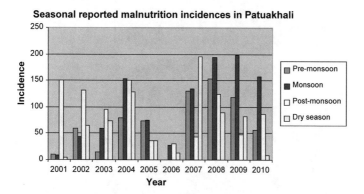

Fig. 9.8 Seasonal occurrence of malnutrition in saline-prone Patuakhali district from 2001 to 2010. *Source* Rahaman (2013)

agricultural activities were hampered and caused unemployment that resulted in food shortages and, consequently, increases in malnutrition. Natural hazards also damaged crops and livestock and increased unemployment rates in the affected area, causing undernourishment in the same period (Fig. 9.8). Skin diseases were highest during the monsoon and post-monsoon seasons, particularly due to the frequency of flooding and waterlogging in the same period (Fig. 9.9).

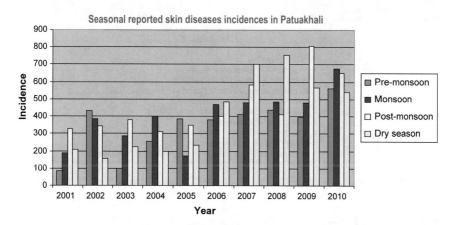

Fig. 9.9 Seasonal occurrence of skin diseases in saline-prone Patuakhali district from 2001 to 2010. *Source* Rahaman (2013)

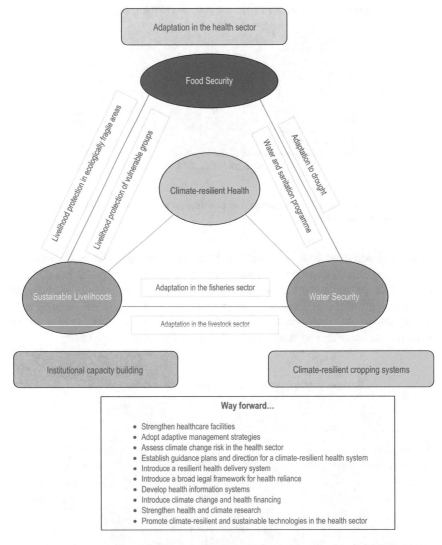

Fig. 9.10 Pathways to climate-resilient health. *Source* The authors based on BCCSAP (2009)

9.3.4 Health Trends in Heat and Cold Waves

Bangladesh also experiences health-related problems as a result of climatic hazards such as heat waves and cold waves each year. Heat waves occur in Bangladesh during the summer, when the maximum temperature goes above 36 °C. Sometimes the minimum temperature goes below 10 °C and a severe cold wave occurs over the western and northern parts of the country (Khatun et al. 2016). Researchers found that these heat and cold waves affected infants and the elderly (aged 60 years and older) the most (Lindeboom et al. 2012). Urban areas were seen to be facing an increased summer mortality peak, particularly regarding cardiovascular mortality (Burkart et al. 2011). A significant increase in deaths in rural Bangladesh was observed due to cardiovascular, respiratory and perinatal causes during periods of low temperature (Hashizume et al. 2009).

9.4 Pathways of a Climate-Resilient Health System

It is necessary to take steps and action to increase health resilience within a changing climate. The steps and actions taken will vary depending on the climate-vulnerable community and location. Several activities are currently in place to support a comprehensive climate-resilient health system in Bangladesh and are being undertaken by different actors, including governmental and non-governmental actors, development partners, private sector representatives and local communities. These actors are working to address different health-related issues and to build health system resilience to climate change by strengthening the health-related infrastructure in Bangladesh.

This section will first explore Bangladesh's institutional health system and its responsibilities. It will then look at the existing healthcare system and explore projects to improve wellbeing in terms of livelihoods, and food and water security, which are central to the support of climate-resilient health pathways.

9.4.1 Institutional Responses to Health Resilience

A strong institutional response is recognised as a vital determinant for building resilience capability in developing countries like Bangladesh. Though a significant attempt has been undertaken to promote resilience of the country's health system, there is still room for improvement at the institutional and administrative levels that could further secure the process (WHO 2015). However, there remains a lack of service centres and educational programmes in medical centres and medical colleges on climate-sensitive diseases. Such centres and programmes are vital to ensuring climate-resilient health delivery systems. From this perspective, a strengthened institutional response is needed to build climate-resilient health systems for Bangladesh. As a result, governmental, non-governmental, inter-governmental and development partner institutions should pay more attention to what is needed for the country's health system to become climate-resilient.

Adaptive management strategies could be an essential adaptation practice for promoting resilient public health systems in Bangladesh. Adaptive management is a management model possessing unique and indispensable components that can enhance the effectiveness of facility restoration in response to management choices and for its dynamic interplay between managers, stakeholders, interventions and policy responses (Whicker et al. 2008; Hess et al. 2012). There are three specific tools for facilitating adaptive management processes: assessment tools, modelling tools and decision support tools. Assessment tools identify and locate hazards and vulnerable populations; modelling tools project or assess appropriate climate-induced health warnings using scenarios; and decision support tools evaluate adaptation options (Hess et al. 2012). Hence, the adaptive management framework is likely to be a helpful strategy for building resilience through strengthening learning at all levels and rebuilding management plans to distinctly climate-sensitive health risk.

Though several institutions are empowered with regulating the behaviour of health service providers and standardising services, regulatory control is highly incorporated within the health ministry. The stewardship role of the public sector has been constrained by a weak legal framework and the institutional inadequacies of administrative functionaries (WHO 2015). Articles 15 and 18 of the Constitution specify the State's duty to provide the basic needs of life (including medical care), raise nutrition levels and improve public health. Along with this broad legal framework, the government has developed several policies and programmes under the *Ministry of Health and Family Welfare* (MoHFW). Table 9.2 summarises Bangladesh's institutional and stakeholder roles for securing a climate-resilient health delivery system.

Table 9.2 Main Contributions of different interested parties and institutions for promoting climate-resilient health systems

Stakeholders	Key Role
National government (Ministries of Environment & Forests and Health & Family Welfare)	• Guides all climate change associated policy issues and represents the country's climate-sensitive health situation at international negotiations under the UNFCCC • Ministries guide Bangladesh's preparation and response to climate-resilient health in the national parliament • Leads the *Local Consultative Group* (LCG) on Climate Change and Health, which plays the highest coordination role among the government and development partners on climate change and health agendas • Active in raising and mobilising funds and introducing mechanisms for dealing with climate change and health
Local government (e.g., Union Parishads)	• Acts as the primary medium of interaction for most of the issues affecting the community people and their livelihoods • Local Government Standing Committees (all committees formed by the UPs for accomplishing its functions efficiently)

(continued)

Table 9.2 (continued)

Stakeholders	Key Role
	perform climate change relevant activities, e.g. disaster management, health, education, agriculture, etc • LGIs partner with NGOs and donors to implement climate-resilient health initiatives at the local level
Individuals, communities and households	• Make the strategies implemented by other stakeholders effective. They are the primary audience, users, and a workforce of health services and health security-related activities • Principal actors in strategising, establishing and developing healthcare systems and other health support activities to ensure their effectiveness, relevance, and sustainability
Academic/research institutes (ICDDR,B; James P Grant School of Public Health)	• Shape the global discussion on equity issues and intergenerational aspects of climate change and health • Research outputs can inform the evaluation of what is already working, what needs to be improved and the potential options for such improvements • Small scale research on health impacts due to climate change can contribute to decision making in the health sector, which is being carried out by different research institutes
Media (Radio; Television; Newspaper)	• Promotes awareness through news coverage, use of readers' forums and seminars • Helps raise national debate on climate-resilient health and financing by bringing the issues to the attention of political leaders, including parliamentarians and development partners • Covers international negotiations and other major events, e.g. Conference of the Parties to the UNFCCC • Issues success and failure stories of climate change projects
Private sector	• Ensures more financial support and collaborate with government, NGOs and health institutes to promote health facilities for climate-vulnerable communities
Development partners and international financial institutions	• Small-scale financing to acquire the knowledge base on climate modification • Financial support to NGOs to integrate climate-resilient health in development plans • Technical and fiscal backing for the government to develop climate-resilient health plans

(continued)

Table 9.2 (continued)

Stakeholders	Key Role
Non-governmental organisations and civil society organisations	• A central link between development partners, government, and communities and, therefore, a vehicle through which activities and strategies can be communicated and implemented • Raising awareness on elementary concepts of climate change and health, causes, risks and potential means of addressing these issues • Advocacy, campaigning, and participatory research on responses to climate-resilient health

Source Adapted from Asia foundation (2012) that granted permission

9.4.2 Climate-Resilient Health Policy

Securing climate-resilient health in Bangladesh is a key focus of the government's vision for responding to climate change. The *Bangladesh Health Policy* (BHP) (2011) aims to establish surveillance for adverse health effects of climate change and create ways to prevent climate change associated health disasters and diseases. It also ensures availability of emergency relief including essential health services, medicines, and other goods to ensure health security for the victims of climatic and natural hazards (BHP 2011). The government recognises health as a priority issue in the *Bangladesh Climate Change Strategy and Action Plan* (BCCSAP), which, drafted in 2008 and updated in 2009 (BCCSAP 2009), was a substantial expansion of the Bangladesh National Adaptation Program of Action (NAPA) (NAPA 2009). The government established the *Climate Change Trust Fund (*CCTF) under the Bangladesh Climate Change Trust (BCCT) in 2010 to provide financial support to projects under the BCCSAP (2009). The government recognises health as a priority issue in the BCCSAP, which highlights six pillars needed to ensure that the poorest and most vulnerable in society are protected from climate change (BCCSAP 2009). The first pillar relates to Food Security, Social Protection, and Health. Under this pillar, the *Government of Bangladesh* (GoB) has developed nine strategic programmes to ensure climate-resilient food security, social protection, and health in Bangladesh. These are:

(a) Institutional capacity for research towards climate-resilient cultivars and their dissemination
(b) Development of climate-resilient cropping systems
(c) Adaptation to drought
(d) Adaptation in the fisheries sector
(e) Adaptation in the livestock sector
(f) Adaptation in the health sector

(g) Water and sanitation programme in climate-vulnerable areas
(h) Livelihood protection in ecologically fragile areas, and
(i) Livelihood protection of vulnerable socio-economic groups (including women)

Though resilience to the adverse impacts of climate change in the health sector has been somewhat acknowledged in both Bangladesh's health policy and the BCCSAP, implementation and monitoring of strategies at the local level is limited. Furthermore, practical knowledge and resources are insufficient to ensure sustainability of the programmes mentioned above.

9.4.3 Health Delivery System and Healthcare Facilities in Bangladesh

The government, private sector, NGOs and donor agencies work together to provide a health system in Bangladesh. The government and the private sector are the primary authorities responsible for rendering health services to people. A comprehensive network and integration of various health facilities extend from the community level to the national level. This network includes primary healthcare centres, community clinics, community health workers, medical colleges, hospitals, maternal and child welfare centres, doctors and nurses, and forms the backbone of Bangladesh's health system. Community clinics provide the bulk of communities' primary healthcare services including maternal and child healthcare, reproductive health and family planning services, immunisations, nutrition education, health education and counselling and first aid, and refer patients to higher-level health centres when necessary. Telemedicine services, introduced by the government in 2011, provide people in rural areas with remote medical advice via real-time video conferencing. ICDDR,B, one of the world's leading global health research institutes, contributes to understanding and resolving severe public health challenges confronting Bangladesh.

Despite these achievements, Bangladesh's health system faces challenges. Health service providers often lack the capacity required to ensure comprehensive healthcare settings and services, such as a lack of funding and members of staff. Moreover, there are no specialist services or units to offer health assistance to the climate-vulnerable populations. The increasing frequency of extreme climatic events, such as floods, cyclones and storm surges, affects the health infrastructure, including hospitals, clinics, diagnostic centres and healthcare facilities, which are used for diagnosing, treating and preventing diseases and illnesses. To overcome these challenges and promote resilience in the overall health system, existing programmes should be strengthened and improved. Access to healthcare facilities must be equal, regardless of geographic location, race, ethnicity, age, gender, social class, culture or ability to pay for the services. Coordination between all parties involved in the health system, including the government, private sector, NGOs and donor agencies is also essential.

9.4.4   Sustainable Livelihoods, Food and Water Security

Sustainable livelihoods, food security, and a clean water supply all lie at the heart of human health. In Bangladesh, adaptation strategies reflect comprehensive responses to climate change to ensure sustainability of these three crucial survival elements through livelihood diversification, crop, fishery and livestock adaptation activities and technological pathways to secure a clean water supply and improved sanitation facilities. The BCCSAP addresses extensive planning and investment to protect the health and livelihoods of poor farmers, with a particular priority on the health and nutrition of women and children, which will ultimately help to ensure food security and combat malnutrition. When combined, these elements can provide climate-resilient health benefits to communities in Bangladesh. Recognising the interrelation between these elements is important. For example, in the context of climate change, sustaining food security will often require diversification of livelihood activities. Many livelihoods in rural Bangladesh are directly linked to natural resources and therefore need a reliable water supply.

The pathways to climate-resilient health in Bangladesh are illustrated in the following diagram:

Livelihoods are one of the social determinants of health. Climate change affects natural resources in Bangladesh, which restricts a secure supply of clean water, food, fodder, shelter, medicine, clothing and the capacity to acquire the above necessities. Most of the population of Bangladesh is fully dependent on nature for their water, food, fodder, medicine, shelter and clothing. Therefore, a shortage of natural resources exerts pressure on the health of climate-vulnerable people (people living in areas vulnerable to adverse climatic hazards including floods, drought, salinity intrusion, etc.), as it results in a lack of balanced and nutritious diet. Livelihood diversification, for example moving from a farming (e.g. crops, live-stock and fisheries) to a non-farming (e.g. day labour, small businesses and services) occupation, is a means of livelihood adaptation. Such activities to ensure livelihood security, therefore, are essential to ensuring climate-resilient health.

Agricultural adaptation strategies are among the most prioritised actions in Bangladesh. Their success has significant positive implications for the health of Bangladesh's population, including direct benefits for reducing rates of malnutrition (Yu et al. 2010). Different actors including development partners, government ministries, international NGOs, and local community-based organisations have facilitated the implementation of a wide range of agricultural adaptation measures. These draw from both local knowledge and modern technology, being adapted as appropriate to the geographic location and socioeconomic conditions of their site of application (Yu et al. 2010).

In addition to these agricultural adaptation strategies, productive fisheries and livestock systems are essential for ensuring climate-resilient health and reducing malnutrition rates in Bangladesh. Salt-tolerant fish species need to be identified for aquaculture in water-logged areas and ponds. Furthermore, technological advancement of traditional fishing boats to include weather warning systems could be a useful adaptation choice for fishers so that they can take shelter before a storm or cyclone evolves. Other possible coping options include protection against development of pond aquaculture, provision of alternative livelihoods for vulnerable fishermen, and the introduction of climate-resilient fodder crops and climate-resilient poultry (i.e., semi-scavengers or slatted housing patterns for poultry) and livestock breeds. In different climate change scenarios, Bangladesh has introduced several climate-resilient interventions for the fisheries and livestock sectors specifically designed to have health outcomes (see Yu et al. 2010). It can be noted that all successful agricultural adaptation activities are likely to have health benefits, even if they do not explicitly target that outcome.

The water and sanitation sector is another crucial area for ensuring climate-resilient health. Consequently, numerous technological pathways are employed in Bangladesh to ensure health by securing safe drinking water. In cities, people mostly get access to water through pipelines and deep tubewells. Rural communities use deep tubewells, shallow tubewells, dug wells, *rainwater harvesting systems* (RWHSs), pond sand filters (PSFs), rivers and canals to access water for drinking, agriculture and other domestic needs. The freshwater sources in the coastal areas – groundwater and surface water reservoirs – are severely affected by salinity intrusion. As a result, people get drinking water from interventions such as PSFs, RWHSs and dupe tubewells. However, the supply of safe water remains inadequate and there is a need to identify a response to these water supply challenges in Bangladesh to ensure climate-resilient health.

Several technologies are already being used in communities in Bangladesh to address these challenges. Yu et al (2010) provide some examples of water technology that have been employed in Bangladesh. Though Bangladesh has introduced a secure water supply scheme, the unavailability of freshwater aquifers at appropriate depths limits its benefits. Better health can be supported through the scaling-up of locally appropriate technologies to ensure sufficient freshwater for drinking and domestic purposes. These adaptive techniques are essential for human wellbeing in climate-vulnerable areas in Bangladesh. They fill the supply gap in areas where the existing water supply scheme is inadequate to match the requirements (Islam et al. 2013). By enabling access to safe drinking water, they provide a pathway to climate-resilient health (Table 9.3).

Table 9.3 Technologies for water resilience in Bangladesh

Technology	Application	Cost-effectiveness and social acceptance
Rain water harvesting systems (RWHSs)	RWHSs in small earthen pots or reservoir tanks provide safe water for drinking and other household activities in drought-prone, flood-prone and saline-prone areas at both the household and community levels	Rainwater harvesting tanks in Bangladesh vary in capacity from 500 litres to 3200 litres, costing from 40 to 110 USD *Acceptance*: Rate of acceptance is the best
Pond sand filters (PSFs)	PSFs enable clean water to be collected from a water source through filtration of the water to remove debris and germs, especially in saline-prone coastal Bangladesh	Installation cost is within the range of 420 to 1600 USD with a capacity to supply water from 4000 to 10,000 litre/day *Acceptance*: Rate of acceptance is good
Desalination panels	This technology can be used to remove salinity and germs from the available water in coastal areas to ensure safe drinking water and overcome health burdens of climate-vulnerable coastal people	Very efficient technology but requires huge installation cost on an average 30,000 to 40,000 USD with a capacity of 10–15,000 litres/day *Acceptance*: Rate of acceptance is moderate due to high installation and maintenance cost

Source The authors

9.5 Conclusions

Climate change is increasing the frequency and magnitude of extreme weather events and creating risks that impact health and healthcare facilities in Bangladesh. People living in flood-prone, drought-prone, and saline-prone areas are the most vulnerable to climate-induced health impacts in Bangladesh (CCC 2009b). The frequency of numerous natural hazards and seasonal disasters is perceived to be the primary cause of prolonged outbreaks of various climate-sensitive diseases, including water-borne and vector-borne diseases, respiratory and cardiovascular disease in Bangladesh. Furthermore, the overall vulnerability of marginal people (including the poor, physically/mentally challenged, ethnic minorities, etc.) to any abovementioned hazardous events is high. In addition, within the marginal populations, the susceptibility of women, children and the elderly are of the utmost in order. Vulnerable poor people are suffering the most from climate-sensitive diseases and insufficient healthcare facilities in the absence of a climate-resilient health system. Healthcare facilities need to assess climate change risks and adopt adaptive management strategies to be resilient. Response activities are increasing, but the guidance, plans and direction for climate-resilient health remain limited. Actors involved in health delivery sectors must take immediate and increased action to work towards a climate-resilient health delivery system. For example, climate-sensitive health policies and plans need to be developed, implemented and coordinated by the MoHFW. This could include setting up service centres and educational curriculums in medical centres and medical colleges to tackle climate-sensitive diseases and ensure a climate-resilient health delivery system.

References

Akter T (2009) *Migration and living conditions in urban slums: implications for food security.* Unnayan Onneshan, Dhaka.

Alam A B M M (2011) Climate change and infectious diseases: global & Bangladesh perspective. *J of Shaheed Suhrawardy Med College* 3(1):1–2.

Alam N, Lindeboom W, Begum D and Streatfield P K (2012) The association of weather and mortality in Bangladesh from 1983–2009. *Glob Health Action* 5:53–60.

Alonso D, Bouma M J and Pascual M (2011) Epidemic malaria and warmer temperatures in recent decades in an East African highland. Proceedings of the Royal Society of London *B: Biological Sciences*, 278(1712):1661–1669.

Asia Foundation (2012) A Situation Analysis of Climate Change Adaptation Initiatives in Bangladesh. Asia Foundation, Dhaka.

Banu S, Guo Y, Hu W, Dale P, Mackenzie J S, Mengersen K and Tong S (2015) Impacts of el niño southern oscillation and indian ocean dipole on dengue incidence in Bangladesh. *Sci Rep* 5:16105.

BCCSAP (2009) Bangladesh climate change strategy and action plan. Ministry of Environment and Forests, Government of the People's Republic of Bangladesh, Dhaka.

Bennett C M and McMichael A J (2010) Non-heat related impacts of climate change on working populations. *Glob Health Act* 3:62.

BHP (2011) *Bangladesh Health Policy.* Ministry of Health and Family Welfare, Government of the People's Republic of Bangladesh, Dhaka.

Burkart K, Khan M H, Krämer A, Breitner S, Schneider A and Endlicher W R (2011) Seasonal variations of all-cause and cause-specific mortality by age, gender, and socioeconomic condition in urban and rural areas of Bangladesh. *Int J Equity Health* 10(1):1.

Butler C D (Ed) (2014) *Climate change and global health.* CABI, Wallingford.

Butler C D and Harley D (2010) Primary, secondary and tertiary effects of eco-climatic change: the medical response. *Postgrad Med J* 86(1014):230–234.

Caminade C, Kovats S, Rocklov J, Tompkins A M, Morse A P, Colón-González F J, Stenlund H, Martens P and Lloyd S J (2014) Impact of climate change on global malaria distribution. *Proceedings of the National Academy of Sciences*, 111(9):3286–3291.

CCC (2009a) *Climate change and health impacts in Bangladesh. Climate Change Cell (CCC)*, Ministry of Environment and Forests, Government of the People's Republic of Bangladesh, Dhaka.

CCC (2009b) *Climate change, gender and vulnerable groups in Bangladesh. Climate Change Cell (CCC)*, Ministry of Environment and Forests, Government of the People's Republic of Bangladesh, Dhaka.

Corvalán C, Hales S and Anthony M (2005) *Ecosystems and human well-being: health synthesis.* World Health Organization, Geneva.

Cox F E G 2010 History of the discovery of the malaria parasites and their vectors. *Parasites & Vectors* 3:5.

Dewan T H (2015) Societal impacts and vulnerability to floods in Bangladesh and Nepal. *Weather and Climate Extremes* 7:36–42.

Dey N C, Alam M S, Sajjan A K, Bhuiyan M A, Ghose L, Ibaraki Y and Karim F (2012) Assessing environmental and health impact of drought in the northwest Bangladesh. *Journal of Environmental Science and Natural Resources* 4(2):89–97.

DG-Health (1997) *Bangladesh health bulletin*. Director General of Health Services, Ministry of Health and Family Welfare, Government of the People's Republic of Bangladesh, Dhaka.

DG-Health (2010) *Bangladesh health bulletin*. Director General of Health Services, Ministry of Health and Family Welfare, Government of the People's Republic of Bangladesh, Dhaka.

FAO (2008) *Climate Change and food security: a framework document*. Food and Agriculture Organization of the United Nations, Rome.

FAO, IFAD and WFP (2015) *The state of food insecurity in the world 2015. Meeting the 2015 international hunger targets: taking stock of uneven progress*. Food and Agriculture Organization of the United Nations, Rome.

Francis A and Maguire R (eds) (2016) *Protection of refugees and displaced persons in the Asia pacific region*. Routledge, New York.

Hashizume M, Wagatsuma Y, Hayashi T, Saha S K, Streatfield K and Yunus M (2009) The effect of temperature on mortality in rural Bangladesh—a population-based time-series study. *Int J Epidemiol* 38(6):1689–1697.

Hasib E and Chathoth P (2016) Health Impact of Climate Change in Bangladesh: A Summary. *Current Urban Studies* 4(01):1.

Hess J J, McDowell J Z and Luber G (2012) Integrating climate change adaptation into public health practice: using adaptive management to increase adaptive capacity and build resilience. *Environ. Health Perspect* 120(2):171.

Hossain M, Kathuria R and Islam I (Eds) (2010) *South Asian economic development*. Routledge, New York.

IPCC (2014) Climate Change 2014: *Impacts, Adaptation, and Vulnerability. Part A: Global and Sectoral Aspects. Contribution of Working Group II to the Fifth Assessment Report of the Intergovernmental Panel on Climate Change*. Cambridge University Press, Cambridge.

Islam A K M S, Murshed S B, Khan M S A and Hasan M A (2014) *Impact of climate change on rainfall intensity in Bangladesh*. Institute of Water and Flood Management (IWFM). Bangladesh University of Engineering and Technology (BUET).

Islam M A, Sakakibara H, Karim M R and Sekine M (2013) Potable water scarcity: options and issues in the coastal areas of Bangladesh. *J Water Health* 11(3):532–542.

Khatun M A, Rashid M B and Hygen H O (2016) *Climate of Bangladesh*. (Rep. No. 08/2016), Bangladesh Meteorological Department and Norwegian Meteorological Institute, Dhaka.

Lindeboom W, Alam N, Begum D and Streatfield K (2012) The association of meteorological factors and mortality in rural Bangladesh, 1983–2009. *Glob Health Action* 5:61–73.

McMichael A J, Haines A, Slooff R and Kovats S R (eds) (1996) *Climate change and human health: an assessment by a task group on behalf of the World Health Organization*. The World Meteorological Organization and the United Nations Environment Programme, WHO, Geneva.

MoHFW (2015) *Malaria national strategic plan 2015–2020*. National Malaria Control Programme (NMCP), Communicable Disease Control Division, Directorate General of Health Services, Government of the People's Republic of Bangladesh, Dhaka.

NAPA (2009) *National adaptation programme of action*. Ministry of Environment and Forests, Government of the People's Republic of Bangladesh, Dhaka.

NIPORT (2015) *Bangladesh Demographic and Health Survey 2014*. National Institute of Population Research and Training, Ministry of Health and Family Welfare, Government of the People's Republic of Bangladesh, Dhaka.

Paaijmans K P, Blanford S, Bell A S, Blanford J I, Read A F and Thomas M B (2010) Influence of climate on malaria transmission depends on daily temperature variation. *Proceedings of the National Academy of Sciences*, 107(34):15135–15139.

Rahaman M A (2013) *Impact of climate change on health in Bangladesh*, Balaka Prakashani, Dhaka.

Rahman A (2008) *Climate change and its impact on health in Bangladesh*. Regional Health Forum 12(1):16–26.

Rashid S F, Gani S and Sarker M (2013) Urban poverty, climate change and health risks for slum dwellers in Bangladesh. In: Shaw R (ed) *Climate Change Adaptation Actions in Bangladesh*, Springer Japan, pp 51–70.

Reid H and Sims A (2007) *Up in Smoke? Asia and the Pacific*. IIED, London.

Saha S (2012) Security implications of climate refugees in urban slums: A case study from Dhaka, Bangladesh. In Scheffran J et al. (ed) *Climate change, human security and violent conflict: challenges for societal stability*, Springer, Heidelberg Berlin, pp 595–611.

Save the Children (2015) *Malnutrition in Bangladesh: Harnessing social protection for the most vulnerable*. Save the Children, UK.

Shovo T E A, Howlader M H and Kumar T (2013) Risk and vulnerability of climate change on coastal people: a study form socio-economic and environmental perspective. *Bangladesh Res Pub J* 8(3):195–202.

Streatfield P K and Karar Z A (2008) Population challenges for Bangladesh in the coming decades. *J Health Popul Nutr* 26:261–272.

Sugden F, de Silva S, Clement F, Maskey-Amatya N, Ramesh V, Philip A and Bharati L (2014) *A framework to understand gender and structural vulnerability to climate change in the Ganges River Basin: lessons from Bangladesh, India and Nepal*. Colombo, Sri Lanka: International Water Management Institute (IWMI). 50p. (IWMI Working Paper 159).

UNICEF (2011) *Children's vulnerability to climate change and disaster impacts in East Asia and the Pacific*. UNICEF, Bangkok.

Whicker J J, Janecky D R and Doerr T B (2008) Adaptive management: a paradigm for remediation of public facilities following a terrorist attack. *Risk Anal* 28(5):1445–1456.

WHO (2013) Nipah Virus outbreak in Bangladesh; at: http://www.searo.who.int/entity/emerging_diseases/links/nipah_virus/en/ (7 Aug 2016).

WHO (2015) *Bangladesh Health System Review*. World Health Organization, Geneva.

WHO (2016) Climate change and human health: Ecosystem goods and services for health; at: http://www.who.int/globalchange/ecosystems/en/ (19 July 2016).

World Bank (2013) *Turn down the heat: climate extremes, regional impacts, and the case for resilience*. A report for the World Bank by the Potsdam Institute for Climate Impact Research and Climate Analytics, World Bank Washington DC.

Yu W, Alam M, Hassan A, Khan A S, Ruane A C, Rosenzweig C, Major D C and Thurlow J (2010) *Climate change risks and food security in Bangladesh*. Earthscan, London.

Chapter 10
Internal Displacement Due to the Impacts of Disaster and Climate Change

Sanjib Kumar Saha and Dilruba Ahmed

Abstract It is evident that the greatest single impact of climate change might be on human migration and displacement: the IPCC predicted migration of 150 million people by 2050. This means that by 2050 one in every 45 people in the world and one in every 7 people in Bangladesh will be displaced by climate change. Against this backdrop, the *Comprehensive Disaster Management Programme* (CDMP II) carried out an in-depth assessment and analysis of the trend and impacts of population displacement due to disasters and climate change in Bangladesh. It is found that about 13% belong to the never displaced category, about 46% belong to the temporarily displaced category, about 29% belong to the in-between temporary and permanent category and about 12% belong to the permanently displaced category. The study found that people living in disaster-prone areas are somewhat more vulnerable and pushed to become displaced either temporarily or permanently. However, current knowledge on the relationship between climate change and migration and displacement of people is still limited. Migration and displacement owing to either natural (e.g. natural hazards) or man-made (e.g. climate change, socio-economic) reasons have a significant impact on people, their livelihoods, the surrounding environment and on the utilisation of resources. Understanding the process of migration, displacement in relation with climate change and disaster is an important topic which needs to be considered at policy as well as implementation level.

Keywords Migration · Displacement · Population · Vulnerability Climate

Sanjib Kumar Saha, Response and Adaptation Management Analyst, CDMP II, UNDP, Corresponding Author, e-mail: kumarsanjib234@gmail.com.

Dr. Dilruba Ahmed, Director, Centre for Environmental and Geographic Information Services (CEGIS).

© Springer Nature Switzerland AG 2019   145
S. Huq et al. (eds.), *Confronting Climate Change in Bangladesh*,
The Anthropocene: Politik—Economics—Society—Science 28,
https://doi.org/10.1007/978-3-030-05237-9_10

10.1   Introduction

The issue of climate change induced migration has received much attention in recent discourse. In 1990, the First Assessment Report of the *Intergovernmental Panel on Climate Change* (IPCC) indicated the significance of the impacts of climate change on human migration (Shamsuddoha/Chowdhury 2009). The report predicted the migration of 150 million people by 2050. More recent studies (i.e. Brown 2008) show an even more terrifying figure of climate change induced migrants: a ten-fold increase in today's population of documented refugees and internally displaced persons. This means that by 2050 one in every 45 people in the world will be displaced by climate change. This figure has a disproportionate spread. Notably, in Bangladesh, one in every 7 people will be displaced by climate change by 2050 (Shamsuddoha/Chowdhury 2009).

Trend analysis of internal displacement is difficult due to limitations in available data. In response to this, the *Comprehensive Disaster Management Programme* (CDMP) conducted an in-depth assessment and analysis of population displacement trends due to disasters and climate change. Four types of climate change induced hazard, namely riverbank erosion, flooding, salinity and waterlogging, were identified as the main catalysts for internal displacement.

Bangladesh experienced heavy rainfall in July and August 2011 (413.8 mm against the monthly average of 332.1). Although it did not cause flooding in other parts of the country, severe waterlogging occurred in three coastal districts: Satkhira, Jessore and Khulna. Satkhira, the most affected district, was inundated by 5–7 feet of water that resulted in the displacement of many people and destruction of houses, standing crops, homestead-based livelihoods and the local market (UNDP 2011).

10.2   Objective of the Study

The overall aim of the research discussed in this chapter was to carry out an in-depth assessment and analysis of the trends and impacts of population displacement due to disasters (riverbank erosion, inundation, salinity and waterlogging). The specific objectives were to:

- Identify population displacement trends in terms of prevalence, incidence and options (up to 2030 or beyond).
- Analyse the social, economic, environmental and other impacts of internal displacement on the directly affected population and the receiving/host communities.

10.3   Research Methodology and Approach

The research was conducted by CDMP jointly with the *Center for Environmental and Geographic Information Services* (CEGIS) in 2013. It adopted a mix of quantitative and qualitative methods. The present situation, parameters and scenarios of climate change and disasters were assessed through a review of available literature. Existing government documents, survey reports and research reports provided the historical trends of internal displacement. A detailed household survey was conducted to explore the vulnerabilities and decision-making processes behind displacement. Focus group discussions, case studies and sharing and brainstorming sessions were also conducted at the local, regional and national levels.

The impact analysis framework, used in this study to analyse climate change impacts, is divided into three different parts (Fig. 10.1). Part A shows the linkages between climate change and hazards; Part B outlines the cause and effect relationship between the hazards and internal displacement (situation analysis); and Part C focuses on the impacts (social, economic, environmental and demographic) of internal displacement on destination/host communities (*denoted by A, B, C and D*).

The study areas were selected on the basis of the intensity of natural hazards occurring in the area (in terms of area coverage). In order to select the study areas, a screening survey examining hazard intensity was conducted in 48 Unions of 24 Upazilas under 8 Districts. Based on the survey findings, 29 Unions of 17 Upazilas under 8 Districts were selected for the study. Within these 8 districts, 926 households were surveyed as part of the study. Figure 10.2 indicates the climatic hazards associated with some of the study areas and the study techniques applied.

10.4   Trend Analysis of Internal Displacement

Based on household responses, this section considers reported experiences of displacement, focusing on the experiences of households that have been displaced either temporarily or permanently. Those interviewed were able to recall several years of displacement trends but the study has considered only the recent past since it is easier for people to recall accurately.

10.4.1   Displacement Due to Floods

Flooding is a very common disaster in Bangladesh, and the country has repeatedly faced severe and devastating floods over the past few decades. Flooding affects up to about 80 percent of the country's land. In a normal year, 20–25 percent of the country is inundated by overflowing rivers and drainage congestion (MoEF 2005). Figure 10.3 highlights the percentage of households in the survey sample that were

A: Linkage between climate change and hazards

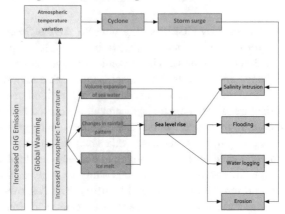

B: Relationship between hazards and displacement

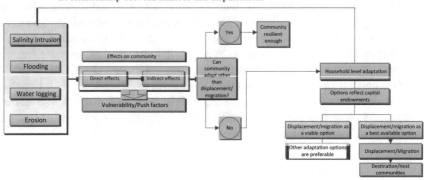

C: Impact of internal displacement on destination/host communities

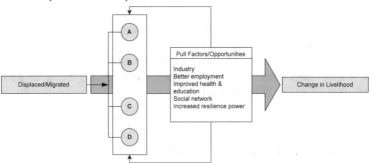

Fig. 10.1 Frameworks for situations surrounding and impact analysis of internal displacement. *Source* The authors

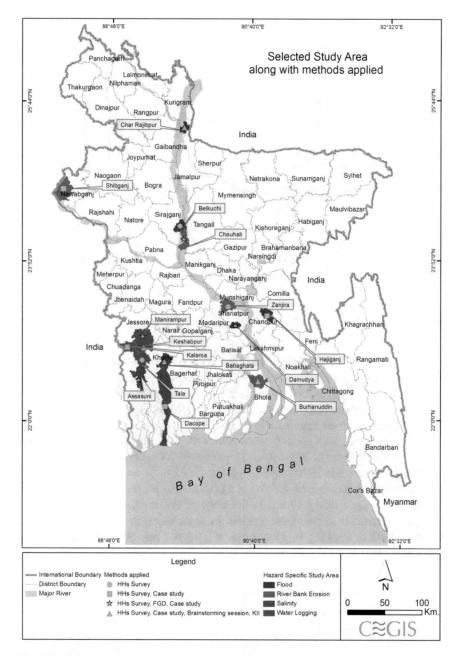

Fig. 10.2 Selected study area and methods applied. *Source* Authors own based on map from CEGIS, Dhaka

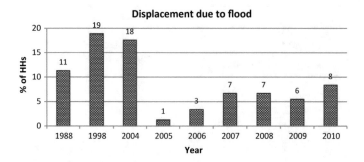

Fig. 10.3 Proportion of displaced households due to floods. The authors

displaced due to floods. About 11 percent recalled being displaced in 1988, about 19 percent in 1998 and about 18 percent in 2004. These years correspond to years of very severe flooding. Almost all of the households were displaced temporarily while only one household was displaced permanently.

10.4.2 Displacement Due to Riverbank Erosion

According to a study by CEGIS (2006), the present rate of riverbank erosion of the Jamuna River is about 2500 ha per year while riverbank erosion of the Padma River is about 1500 ha per year. In 2008, erosion along the banks of the Jamuna was 530 ha, the Ganges around 880 ha, and the Padma 535 ha, of which about 85, 75 ha, and about 100 ha respectively contained settlements (DoE 2012). In analysing trends, it was found that, of the survey sample, about 19 percent were displaced due to riverbank erosion in 1998, about 22 percent in 2005, about 15 percent in 2008, about 11 percent in 2009 and about 13 percent in 2010 (see Fig. 10.4).

It is evident that riverbank erosion is a disaster that pushes people toward rapid displacement. Case Study 10.1 tells the story of Sabura Khatun's experience of being displaced due to riverbank erosion. In some cases, flooding aggravated the situation further, resulting in a comparatively higher rate of displacement in the years when severe floods occurred.

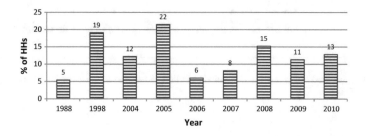

Fig. 10.4 Proportion of displaced households due to riverbank erosion. *Source* The authors

Case Study 10.1: Miseries created by nature. *Source*: This box was produced by authors of the migration chapter. Who granted permission to include the text here.

Sabura Khatun (age 27)
Sabujbagh, Char Razibpur, Kurigram
Affected by Riverbank Erosion

Sabura Khatun

I am the seventh of the 10 children of my parents. My father was a share cropper who was always struggling to maintain such a big family. He sometimes worked as an agricultural day labourer to supplement the little he earned from share cropping. In 2006, I was given in marriage to my now-divorced husband, Shafiqul. We both came from the same village of Modoner Char and, like my family, he had settled here after being displaced due to a natural disaster. A year after we got married we were blessed with a son who brought great joy to our lives. However, later that year, our happiness was destroyed forever by riverbank erosion when it engulfed our house and that of my parents. That was the beginning of a very difficult time for us, as we did not have enough money to rebuild our home. My husband demanded that I ask my father to get him tin sheets to build the roof of a new house, but my father was also in financial crisis with no means of buying materials for their own home. He requested my husband to give him some time, but in the end he could not arrange the money.

Sabura Khatun with her parents

Shafiqul berated me about it and it was not long before he beat me for failing to come up with the money. My situation became intolerable and then one day, he decided to divorce me. After separation, he went to Dhaka and I came back to my father's house. In our society, no woman can live with her parents after marriage, so it was really difficult for me to live in my father's house for long. I realised that there was nobody on earth who would stand beside me and that I had to do everything on my own for survival. I had to think of my son, as my only duty now was to help him become established in life. I managed to get a job here as a cook at the upazila dormitory. I earn 2000 taka per month and live in a small house that I rent. I earn very little and so there are periods when my son and I have only one or two meals a day. I cannot even buy new clothes for my son but have to wait for occasions such as Eid ul Fitr when poor people like me receive clothes in zakat. I am always afraid of what will happen to me and my son if I ever lose my job. It is the only support I have and without it there will be no one to give us shelter.

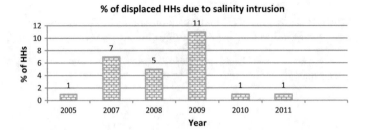

Fig. 10.5 Proportion of displaced households (Hhs) due to salinity intrusion. *Source* The authors

10.4.3 Displacement Due to Salinity

Displacement due to salinity does not occur frequently; it occurs when drastic natural events take place. Analysis shows high displacement rates due to salinity intrusion in 2007 and 2009 (Fig. 10.5). Two natural disasters occurred in those years – cyclone *Sidr* in 2007 and cyclone *Aila* in 2009. Both destroyed a significant amount of infrastructure including embankments, road networks, institutions etc. and eventually pushed saline water into the mainland. These cyclones and consequent salinity also damaged crops and vegetation, destroyed livestock, caused a number of fatalities, and resulted in many people becoming displaced between 2007 and 2009.

Scientists predict that more areas will come under tidal influence due to sea level rise. This would result in an increase in salinity levels and intrusion in coastal areas such as the study areas within Khulna and Satkhira districts. The areas are under high salinity influence, which is projected to increase by 5 to 15 parts per thousand by 2050 (CEGIS and IWM 2007).

10.4.4 Displacement Due to Waterlogging

Waterlogging is a result of a combination of factors, including excessive monsoon rains, inadequate drainage, mismanagement and a lack of maintenance of embankments, increased sediment and siltation of rivers, restricted river flows due to embankments built for shrimp farming, and the release of water from barrages upstream (Roy 2011). Bangladesh experienced heavy rainfall in July and August 2011 (413.8 mm against the monthly average of 332.1 mm) and, though it did not cause flooding in other parts of the country, extensive waterlogging was created in the three coastal districts of the study, namely Satkhira, Jessore and Khulna. Satkhira, the most affected area, was inundated by 5–7 feet of water that caused massive displacement in the population and severe destruction of houses, standing crops, livelihoods and local markets (Fig. 10.6).

Fig. 10.6 Proportion of displaced households due to waterlogging. *Source* The authors

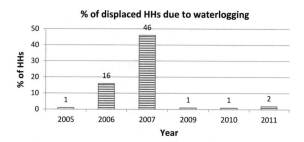

In an overall assessment, no displacement patterns were found over the designated time period. However, it was found that there was much waterlogging in 2006 and 2007. People stated that cyclone Aila contributed significantly to displacement in 2007. Waterlogging is a slow onset disaster. It is not considered as a single factor contributing to displacement, as other factors such as flooding may contribute. In the case of waterlogging, displacement takes place very slowly and is usually temporary, but the duration of the displacement is longer than for flooding.

10.5   Impact Analysis

This impact analysis examines the impacts of various social, economic, and other factors associated with selected disasters, on internally displaced populations. It explores these impacts on displaced populations in both their places of origin and in their host communities (i.e. the communities that they move into). There are numerous impacts of internal displacement that influence the lives and livelihoods of both communities. While in many ways these impacts have negative connotations, there is a positive effect of displacement, suggesting that it can be considered as a coping strategy for climate-affected households. Exemplifying this, Case Study 10.2 provides a picture of the benefits that Monoranjan Mondal has experienced and of the opportunities he has created as a result of his household's displacement.

Households in the places of origin and destination are both captured in this analysis. Temporarily and permanently displaced households are considered in unison as 'displaced households'. Comparative analysis is made between the displaced (809) and never displaced (117) households. Education, health, occupation, income-expenditure and assets) are identified as indicators for the impact analysis.

10.5.1   Impact on Education

People that have never been displaced were found to have a better educational status than displaced people. With regards to the displaced households, most children dropped out of education at the outset of their displacement and a number

Table 10.1 Education status by nature of displacement. *Source* The authors

Type of education	Educational status of 7+ aged people (%)	
	Never displaced	Displaced
Not passed class I	31	33
Class I–V	36	31
Class VI–IX	22	18
SSC, HSC or equivalent	4	3
Bachelor degree & above	3	1

of them attempted to get involved in income generating activities for their survival. Children that were never displaced continued their studies in the same institutions in the place of origin where those institutions provided some support to the vulnerable children for continuation of their education (Table 10.1).

People who permanently left their place of origin tried to adjust to a new community and to ensure their children were as educated as others in their new community. However, it was more difficult for temporary migrants, as they could not enrol their children in new institutions because they knew that they had to return to their place of origin. Consequently, their children could not adapt to their new, temporary community.

10.5.2 Impact on Health

The health profile of the households in the study sample indicates that a high rate of people suffered from diarrhoea and fever/cold/cough compared to the national level (see Table 10.2). The incidence of illness among both displaced and never displaced households are similar for all ailments except malaria, which is significantly lower in displaced households. During disasters, especially flooding and waterlogging, people suffered from waterborne diseases owing to the poor sanitation systems making the environment highly unhygienic.

Table 10.2 Ailments of the households

Ailment	Households suffering from ailments (%)		
	Never displaced	Displaced	National
Arsenicosis	2	0.55	–
Diabetes	3	5	5.4
Diarrhoea	64	72	5.08
Fever/cold/cough	99	98	58.03
Malaria	11	3	–
Pox	15	13	0.17
Tuberculosis	2	2	0.18
Others	12	21	6.29

Source The authors

Table 10.3 Occupation of the displaced and never displaced households

Occupation type	Involvement of occupants (%)	
	Never displaced	Displaced
Agriculture	20	20
Production & related work and transport work	20	23
Service work	7	4
Retired	35	38
Other	18	15

Source The authors

10.5.3 Impact on Occupation

Displaced households often have to move to places that are far from where they work. This increases their commuting time to work and many of them lose their jobs for failing to follow working schedules. Furthermore, disaster prone places are found to be unfavourable for employment, especially for wage earners, meaning that displaced wage earners tend to seek employment elsewhere. This drives relocation and displacement (Table 10.3).

The study found that households in disaster prone areas increasingly lost interest in agricultural activities due to the loss of cropland. It many cases these households have started to involve themselves in small businesses and transport related work and services.

10.5.4 Impact on Income and Expenditure

The average household annual income and expenditure and the resettlement cost of each household for respective disasters (if occurred) is presented in Table 10.4. It was found that the erosion-affected households spent the highest amount on house repair and resettlement, as most parts of their homesteads had been washed away by riverbank erosion. These expenses were lowest in salinity-affected households

Table 10.4 Households (HHs) average resettlement cost compared to yearly income and expenditure

Types of hazard	HHs avg. yearly income (BDT)	HHs avg. yearly expenditure (BDT)	Resettlement cost for each HHs (BDT)
Flood	137539	114750	40570
Riverbank erosion	162411	150552	295132
Salinity	139793	124161	13795
Waterlogging	102122	108940	36784

Source The authors

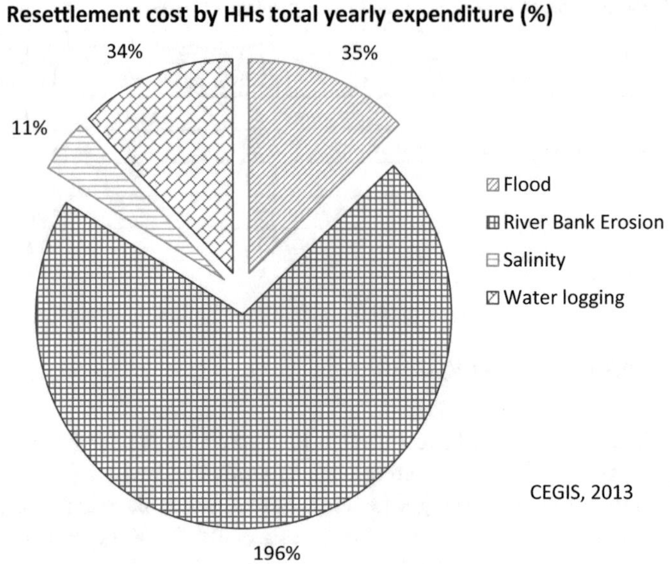

Resettlement cost by HHs total yearly expenditure (%)

Flood
River Bank Erosion
Salinity
Water logging

CEGIS, 2013

Fig. 10.7 Resettlement cost of households (Hhs) by total yearly expenditure for each hazard. *Source* Authors own based on data from CEGIS (2013)

because people only needed minor house repairs and sometimes cut back on their agricultural production.

Figure 10.7 shows the household resettlement costs based on different types of hazard, as a percentage of households' annual expenditure. Households affected by riverbank erosion spent, on average, almost twice (196%) their annual expenditure on resettlement after each disaster. Expenditure is particularly high for erosion-affected households, as most of the time they have to rebuild their homes in new places. As a result, they have to borrow money, which becomes a burden in the long run. For the remaining disasters, households could repair their existing houses afterwards, so their resettlement cost is much lower than that of the riverbank eroded households.

10.5.5 Impact on Assets

Assets such as poultry, livestock and gardens are endangered by the four disaster types. Most people surveyed had taken loans to acquire these assets, so had to spend most of their savings after losing these assets. As a result, households that had lost assets tried to permanently migrate and settle in new, safer areas.

10.5.6   Analysis to Identify Impact on Demographic Indicators

The impact on the above demographic features was statistically verified using a logistic regression test to determine the significance of different variables. Here, the comparison or dependency is measured for two types of households: (i) never displaced households and (ii) temporarily displaced households. Never displaced households are used as the reference category and the two types of households are identified by probability sampling.

As Table 10.5 shows, the dependency test suggested that the level of displacement decreases with the increasing land-holding size. This is with the exception of marginal land-holding households who are slightly more commonly displaced than small land-holding households. Therefore, the large land-holding households have the lowest chance of displacement.

The test also showed that the likelihood of household displacement decreases with increasing family size. The analysis indicates that displacement is lowest for the large family size households (0.826), the second lowest for the medium size households and the highest for the small family size households (Table 10.5).

The chance of household displacement seems to decrease with the increasing level of academic education of the household head. Our analysis found that households with heads having secondary or above education are less likely to be displaced than those with heads who are illiterate or only have primary education.

Table 10.5 Results of logistic regression analysis

Dependent variable	Independent variable	Categories of independent variable	B[a]	Significance[b]	Exp (B)[c]
Temporarily displaced	Intercept	Intercept	3.162	0.000	
	Land holding category	Marginal	0.075	0.871	1.078
		Large	−1.329	0.003	0.265
		Small	**Reference**		
	Family size	Medium	−0.003	0.993	0.997
		Large	−0.191	0.642	0.826
		Small	**Reference**		
	Education status of HHH	Primary	−0.243	0.425	0.785
		Secondary and over	−0.305	0.377	0.737
		Illiterate	**Reference**		
	Primary occupation of HHH	Labor	0.023	0.949	1.023
		Others	0.258	0.443	1.294
		Agriculture	**Reference**		
	Types of Hazard	Flood	−2.557	0.000	0.078
		River Bank Erosion	1.920	0.019	6.822
		Salinity	0.622	0.338	1.862
		Waterlogging	**Reference**		

*Note The **reference** category is: Never displaced*
[a]Coefficient of regression analysis
[b]P value; probability of rejecting null hypothesis
[c]Odds ratio; (Temporary displaced/Never displaced)

This is verified by the Exponential Intercept Value (Exp (B)) of 0.737, which indicates that displacement is lower for HHHs having secondary or higher education than for illiterate HHHs.

On the other hand, it is observed that the chance of displacement for households headed by labourers or other professions is higher than the households headed by farmers. In terms of Exp (B), other classes of households have more chance of displacement (Exp (B) equal to 1.294) than households headed by farmers and labourers (see Table 10.5).

Among the four different types of disasters examined, the chance of displacement is most significant for riverbank erosion. The likelihood of displacement is higher (Exp (B) at 5.466) for households affected by riverbank erosion than for the households affected by waterlogging (see Table 10.5).

Case Study 10.2: Fighting adversity to build a better life! This was produced by authors of the migration chapter who granted permission.

Monoranjan Mondal (age 49)
Education: H.S.C. Profession: Farmer
Affected by Waterlogging

I am the second son of my father Mr. Bipod Vonjan Mondal of Sujatpur village in 15 Kultia union of Monirampur upazila under Jessore district. In spite of being a small farmer, my father used to maintain our family very well. Natural disasters, however, snatched away the easy life we had. Waterlogging is the main form of hazard in our village, which first struck in the early 80 s. Since then, the situation has deteriorated. The village has been surrounded by water continuously over the last 8 years. Sujatpur is mainly a rice cultivated area where waterlogging for 7–8 months hampers rice production and causes untold sufferings to nearly 100 households. Crop production by marginal and small farmers is seriously hampered. Due to crop losses and no alternative sources of income, our family fell into serious financial crisis. I had 126 units of agriculture land which remained fallow due to waterlogging. 10 decimal of homestead land was also blocked by water all year round. Road communication became difficult as it was also disrupted by waterlogging. The wall around

Monoranjan Mondal

Monoranjan shows the level of water

Interview with Monoranjan

our house broke due to wave action during the
monsoon and every year I had to spend around
10000 taka to repair my house. Insufficient
work opportunities and all the other factors
finally forced me to move to the village
Sujatpur Uttar para of the same union in 2005.
I prefer this area because it is more elevated
than my previous village. Moreover, this vil-

Monoranjan on his way to the Village Market

lage is near my old home and people of my religion live here. As this area is
free from waterlogging, it is very easy to look after my land and homestead.
I have been able to adjust easily with the host community because of my
education, knowledge and attitude. Since migrating to this area, I have been
living a good life free from tension. I get more than 3 tons of food-grain every
year from my land and so I am able to look after my family very well.

I have built a house on 3 units of land and started crop cultivation in a
share cropping arrangement. I also have a seasonal business along with cow
rearing and a small poultry rearing initiative. I had discontinued my studies in
the early 90 s to help my father by earning an income for our survival. It was
my special desire to continue my education to get a good job, but my dream
did not come true because of what my family suffered due to natural disasters.
However, I remain determined to fulfil my dream through my children. I have
ensured that unlike me, my children have the opportunity for education. My
eldest daughter is a college graduate and the younger one has obtained
Masters of Arts (M.A.) and is now working as a teacher. My only son is an
Honours student in Dhaka College. I am waiting for the successful completion
of his studies as it will be the fulfilment of all my dreams.

10.5.7   Migration and Displacement Management Strategies

Findings and observations from the study reveal that there are enormous challenges
caused by displacement due to various impacts of climate change and disaster.
Those who are forced to leave their ancestral homes take shelter in places where
they neither have rights nor are provided with basic social services. However, as the
study finds, there are examples where people have settled in new locations,
leveraging the opportunities to live better lives. So, while displacement can leave
people in a state of misery, it also offers opportunities for coping and provides
pathways to a resilient future.

In the context of Bangladesh, the issue of displacement and migration has to be
addressed at both the international and national levels. There is also a need to
incorporate it at the local level, so that proper action can be taken to deal with
related problems. A national strategy encompassing the rights and entitlements of

the displaced people could possibly be sufficient to deal with the emerging problem. The strategic responses cannot simply be assumed to be a neutral activity affecting everyone equally and in a positive way. Any viable approach to displacement should be based on the principles of human rights enshrined in the international and regional human rights instruments.

Based on the findings and results, the following recommendations are made:

- Mainstream the issue of migration and displacement into national development plans;
- Develop policies to ensure social protection of more vulnerable or poorer migrants;
- Devise a contingency plan for relocation of communities threatened by river erosion;
- Design and implement community-specific local level projects to help migrants and displaced people to cope with the changing environment.

References

Brown O 2008. The Numbers Game. *Forced Migration Review*. October 2008, Issue 31.

CEGIS 2006. *Predicting Riverbank Erosion along the Ganges*. Center for Environmental and Geographic Information Services (CEGIS), Dhaka, Bangladesh.

CEGIS and IWM 2007. *Investigating the Impact of Relative Sea-Level Rise on Coastal Communities and their Livelihoods in Bangladesh*. Center for Environmental and Geographic Information Services (CEGIS), Dhaka, Bangladesh.

CEGIS 2013. GIS data, Center for Environmental and Geographic Information Services (CEGIS), Dhaka, Bangladesh.

DoE 2012. *Second National Communication (SNC) of Bangladesh for the UNFCCC*, Department of Environment, Ministry of Environment and Forests, Dhaka, Bangladesh.

Ministry of Environment and Forest (MoEF) 2005. *National Adaptation Programme of Action (NAPA) [Report]*. - Dhaka : Ministry of Environment and Forest (MoEF), Government of the People's Republic of Bangladesh (GoB), MoEF; and United Nations Development Programme (UNDP), November; at: http://unfccc.int/resource/docs/napa/ban01.pdf.

Roy CD 2011. *Vulnerability and population displacements due to climate-induced disasters in coastal Bangladesh*, Centre for Geoinformatics, University of Salzburg, Austria.

Shamsuddoha M and Chowdhury R K 2009. *Climate Change Induced Forced Migrants: in need of a dignified recognition under a new Protocol*, Equity and Justice Working Group Bangladesh.

UNDP 2011. *Waterlogging in Satkhira District: An Analysis of Gaps between Needs and Response, Early Recovery Facility*, UNDP Bangladesh.

Chapter 11
Climate Change Adaptation in Urban Areas: A Critical Assessment of the Structural and Non-structural Flood Protection Measures in Dhaka

Md Asif Rahman and Sadya Islam

Abstract Dhaka, the capital city of Bangladesh, is susceptible to floods because of its topography, large population, and inadequate infrastructure. Almost half of Dhaka's population live in the low-lying and flood-prone areas, and the local communities suffer from economic stress due to flooding as well as having an absence of flood protection measures in these areas. The economic damages caused by flooding have a severe impact on national economic development and growth. Unplanned urbanization intensifies damage from flood incidents, including caused by the migration into Dhaka from rural areas as people move to the city in search of jobs and to recover from the impacts of natural disasters like cyclones, droughts, and river erosion in rural areas. This chapter outlines Dhaka's current situation with regards to flood exposure and risks from climate change as well as summarising key measures to reduce flood risk in recent years.

Keywords Urban · Climate · Protection · Dhaka · Flooding · Exposure Risk

11.1 Introduction

Bangladesh, due to its vulnerable geographical location, high population density, poor infrastructure quality, and unstable economic and political situation, is considered one of the most vulnerable countries in the world to climate change and climate-induced hazards. Among the climatic perils, flooding (because of its extent and damage potential) poses a significant threat to people. Dhaka, the capital city, is particularly susceptible to floods because of its topography, large and growing population, and inadequate infrastructure. Dhaka is also the center of Bangladesh's

MD Asif Rahman, Department of Geographical and Sustainability Sciences, University of Iowa, Iowa City, USA, Corresponding Author, e-mail: asif0236@gmail.com.

Sadya Islam, School of Urban and Regional Planning, University of Iowa, Iowa City, USA.

© Springer Nature Switzerland AG 2019   161
S. Huq et al. (eds.), *Confronting Climate Change in Bangladesh*,
The Anthropocene: Politik—Economics—Society—Science 28,
https://doi.org/10.1007/978-3-030-05237-9_11

administrative, educational, economic and social activities, and any damage and impediment to the city's functions can threaten the country's overall economy. Therefore, this chapter reviews and evaluates flood protection measures for safe-guarding Dhaka. Firstly, it outlines Dhaka's current situation concerning flood exposure and risk from climate change. Secondly, it summarizes several measures to reduce flood risk and damage undertaken in recent years, and thirdly, it evaluates the performance of these actions.

Dhaka is surrounded by various water bodies, namely the Buriganga River to the south, Balu River to the east, Tongi Khal to the north and the Turag River to the west (see Huq and Alam 2013). These rivers are distributaries of three major rivers, the Ganges, Brahmaputra, and Meghna, which form the flat deltaic plain. Dhaka is at the center of this flat plain (Bala et al. 2009). The Dhaka Metropolitan (DMP) area is approximately 304.16 km^2, and the total population of the area is around 14 million (Dewan 2013). The city is divided into Dhaka East and Dhaka West. Embankments surround the western part, but dam construction in the flood plains of the east remains under consideration (Dasgupta et al. 2015).

Almost half of Dhaka's population live in the low-lying and flood-prone areas, and the local communities suffer from severe physical, social and economic stress due to flooding, water logging, and drainage congestion. The absence of flood protection measures makes the eastern part more flood-prone, and the whole city vulnerable to flood hazards (Haque et al. 2010). Flooding also causes severe environmental problems such as contamination of drinking water, riverbank erosion, and mosquito infestation.

Dhaka has experienced several flood incidents due to the overflow of the encircling rivers. This excess runoff is exacerbated by unmanageable upstream flow, the unplanned and congested city fabric and unpredictable climate. The city has already experienced significant floods in 1954, 1955, 1970, 1974, 1980, 1987, 1988, 1998, 2004 and 2007. The 1998 and 2004 floods were catastrophic, taking into account the extent of affected areas and duration of waterlogging. In both years, a significant portion of the city was inundated, and more than 50% of city dwellers were severely affected. The impacts were higher among the poorest segment of the population, including day-laborers, rickshaw pullers, slum dwellers and the homeless.

The total economic cost of the floods in 1988, 1998 and 2004 was US$11.16 million, US$4.4 million and US$5.6 million respectively (Dewan 2010; Gain et al. 2015). Being a developing country where almost one-third of the population live below the poverty line, such huge financial loss has implications for economic growth and development progress. According to several studies, Dhaka is still under severe threat from urban flooding due to the impacts of climate change (Bala et al. 2009; Gain et al. 2015; Dewan 2013). Flood mitigation measures are therefore essential to address the economic, social and environmental impacts of urban flooding.

11.2 Flood Risk in Dhaka

According to the framework used by Gain et al. (2015), the overall flood risk of the city depends on three factors: the intensity and frequency of the "hazard", the "exposure" of the elements such as people, infrastructure, agriculture and others and the "vulnerability" of the exposed elements (see Gain et al. (2015) and Mojtahed et al. (2013) for flood risk assessment techniques in urban areas). Risk can be understood as a result of the interaction between hazard, exposure, and vulnerability (IPCC 2014). A hazard is the potential of a natural or human-induced event to occur that could result in loss and damage to assets such as infrastructure, property, livelihoods, provision of services and environmental resources in addition to injury, loss of life or other adverse effects on health (IPCC 2014). "Exposure" refers to the presence of resources, assets, livelihoods or people in areas or settings that could be adversely affected by a hazard (IPCC 2014). Finally, vulnerability, as defined by the IPCC (2014), is the propensity to be adversely affected, including the "sensitivity to harm", by the exposed elements and the "capacity to cope and adapt".

The high intensity of flood hazard in Dhaka depends on two factors. First, the overflow from surrounding rivers results in flooding almost every year. The upstream flows from improperly managed trans-boundary water sources escalate this problem (Mozumdar 2005). The second factor is the extreme rainfall events that have increased in frequency and magnitude over the last decade due to the effects of climate change (Murshed et al. 2011). Dhaka receives around 2000 mm of rainfall each year, and 80% of this occurs in the monsoon season, thus increasing the probability of flooding (Dewan 2013). Rain events have become unpredictable and erratic due to anthropogenic climate change that can increase the intensity and frequency of hazards, thus heightening overall flood risk in Dhaka (Alam and Rabbani 2007).

The unplanned urbanization across the city also intensifies damage from flood incidents. Massive rural to urban migration by people in search of jobs and to recover from the impacts of natural disasters like cyclones, droughts, and river erosion has increased the demand for housing and essential services in the city. Each year around 500,000 people migrate to and settle in Dhaka city (Cities Alliance 2016). Many illegal slum and squatter settlements are constructed to create accommodation for this growing population, and it is done in an unplanned way, bypassing planning regulations, which are facilitated by local political leaders and middlemen (Morshed 2013; Kabir/Parolin 2012). According to Dewan/Corner (2012), the total urban land area of Dhaka city has increased 67% from the 1990 levels (from around 116.20 km^2 in 1990 to 194.07 km^2 in 2011). Griffiths et al. (2010) also demonstrated the rapid urbanization of the city with most growth occurring in the unprotected eastern part where, due to the lack of an embankment, flooding happens more frequently. The study also predicted that migration towards Dhaka would increase due to the impacts of climate change. Problematically, the greater the migration into the city, the greater the exposure to flood risks, especially for low-income residents living on marginal, flood-prone land.

The third factor that increases the risk of flood impacts in Dhaka is the unsafe conditions of infrastructure. The infrastructure facilities (such as roads, drainage systems, buildings, etc.) in most of the city are inadequate and prone to damage from flood incidents (Braun/Aßheuer 2011). Also, the high population density of the city increases the probability of harm from flood events. Approximately 40% of the total population in Dhaka live in slums that are built in flood-prone areas (Dewan 2013) with very basic infrastructure and unhygienic living conditions (Rashid et al. 2007). The poor socio-economic conditions of most of the city dwellers reduce their coping capacity to flood events and result in damage to their shelters, health, and economic conditions.

The overall flood risk in Dhaka is expected to increase, due to the effects of climate change. It is anticipated that the frequency of extreme rainfall events in Dhaka city could increase 16% by 2050 (Dasgupta et al. 2015). Another analysis also showed that return period of extreme flood events is lessening (which means that extreme events are becoming more frequent) (Dasgupta et al. 2015). The planned improvements to the drainage system of Dhaka city may still lack the capacity to handle such large rainfall events and contribute to the overall flood risk.

11.3 Stakeholders and Measures of Flood Risk Management

At present, the management of flood control measures is the combined responsibility of several organizations (Table 11.1), which might create a jurisdictional impediment. In the past, lack of coordination between stakeholders resulted in a bad management of the culverts and regulators during the 1998 flood (Bala et al. 2009). Dewan (2013) noted that organizations tend to blame each other for critical failures, and are sometimes unwilling to take responsibility for the damage. However, during the 2004 flood, Bala et al. (2009) observed that the level of coordination improved and, as a result, the flood situation was managed more efficiently.

Table 11.1 Stakeholders in flood risk management in Dhaka

Organizations	Responsibility
Ministry of finance	Allocating the budget for flood control measures
Dhaka South City Corporation (DSCC) and Dhaka North City Corporation (DNCC)	Controlling the smaller underground and surface drains within their jurisdiction area
Dhaka Water and Sewerage Agency (DWASA) and Bangladesh Water Development Board (BWDB)	Operating and maintaining the embankments around the city Looking after the larger open drainage canals and pipes
Capital Development Authority (RAJUK)	Installing underground drainage lines during road construction

Source The authors

Existing methods for flood mitigation are traditionally divided into two broad groups: structural and non-structural measures (Dasgupta et al. 2015). Structural measures, according to the definition of the United Nations International Strategy for Disaster Reduction (UNISDR 2015), include physical constructions that aim to mitigate potential hazardous impacts or the application of engineering techniques to achieve hazard-resistance and resilience in structures or systems. In contrast, non-structural measures are those that do not involve physical construction but use knowledge, practice or agreement to reduce risks and impacts, in particular through 'policies and laws, public awareness raising, training and education' (UNISDR 2015).

Andjelkovic (2001) views structural measures as the more conventional method of protection in the form of dams, storage reservoirs, dikes, floodwalls, flood diversion, channels and land treatment practice. Non-structural methods, on the other hand, include 'preparedness, response, legislature, financing, environmental impact assessment, reconstruction and rehabilitation planning' and can contribute directly towards reverting the damage potential of flooding by giving people more strength for post-disaster recovery (Andjelkovic 2001).

11.4 Structural Measures

Buckland Bund was the first ever embankment constructed in 1864 on the northern bank of Buriganga River to protect Dhaka from flooding and river erosion. The need for flood mitigation measures arose after the 1954 and 1955 disastrous floods in the country (Huq and Alam 2003). As a result of these floods, the Master Plan of 1964 was prepared and some significant flood mitigation studies were conducted (Mozumder 2005). While plans for flood protection were under consideration for Dhaka, the pressing need for immediate action arose after the 1987 and 1988 floods. At this time, the Government of Bangladesh initiated a two-phased immediate flood protection and drainage project, "Greater Dhaka Flood Protection Project (GDFPP)", that included enclosing the DMP area with flood embankments, reinforced concrete walls, and providing drainage and flood regulation structures such as sluices and pumping stations. Phase-I was completed in 1992; with a 136 km dam surrounding the western part of Dhaka city (Chowdhury 2003). Alam and Rabbani (2007) identified the following key components of the flood protection measures:

- Approximately 30 km of earthen embankment along the Tongi canal and the Turag and Buriganga rivers
- About 37 km of raised roads and floodwalls
- A total of 11 regulators along the embankment at the outfall of khals (canals) to the surrounding rivers
- One regulator and 12 sluice gates on the khals at the crossings with the Biswa, DIT, Pragati Sarani and Mymensingh roads and the railway line at Uttar Khan

- One pumping station at the outfall of the Kallyanpur khal into the Turag River and another at the outfall of the Dholai khal to the Buriganga River. These pumping stations are for draining rainwater from parts of Dhaka West, and
- A special 10.53 km embankment surrounding the Hazrat Shahjalal International Airport.

Phase-II is yet to start but will surround the city from the eastern fringe with a rail/road embankment that will run for 29 km along the Balu River. Though these flood control and drainage works saved Dhaka west from flooding to some extent, during the 1998 flood, the embankment could not serve its intended purpose due to improper planning. Its height was not enough to contest a flood like the one in 1998. Furthermore, some open culverts[1] and unclosed regulators of the embankment facilitated water intrusion from outside. The embankment worked well during the 2004 and 2007 floods, but the condition of the protection bunds has degraded as settlements have increased and encroached onto them (Bala et al. 2009).

Subsequently, the embankment has brought about significant social, land use and environmental changes in the surrounding area. Agriculture lands in the fringe have been converted to residential areas. Dumping of solid waste and agricultural residue disrupts the area's natural beauty. Groundwater quantity and quality have also been reduced to a great extent. Low-income residents are also being forced to build squatter and slum settlements in these areas, exacerbating flooding problems outside and inside the city by impeding the existing drainage system.

During the monsoon, the higher water level of the rivers reduces drainage capacity, reducing their ability to accommodate the city's surface runoff (Dewan 2013). At the same time, continuous rainfall increases the surface water runoff in the city, and demand for surplus rainwater drainage increases. The reduced water bearing capacities of the river along with the inadequate drainage infrastructure creates internal flooding and waterlogging. To alleviate the problem, DWASA undertook a storm water drainage improvement plan that ensured the proper flow of the city's water bodies by constructing concrete box culverts (Huq/Alam 2003). Another measure was the banning of polyethylene bags to reduce drainage system congestion (Alam/Rabbani 2007). Dhaka has approximately 40 canals and some ponds that act as retention and detention areas. The canals are linked with the river network and create a drainage system. Many of these canals and lakes are no longer effective due to illegal encroachment or acquisition for the construction of buildings (Bala et al. 2009). Mahmud et al. (2011) show the spatial extent of wetlands in Dhaka city over last few decades, with a significant decrease between 1988 and 2009. The loss of wetlands impedes the natural drainage system of the city and thus increases flood risk.

There are 3 large and 66 small pumping stations to reduce inland water congestion; those functioned well until the 2007 flood. Additionally, there are 185 km of drainage pipes under the city to drain storm water (Bala et al. 2009). Dhaka can sustain medium sized rainfall events (200–250 mm/day), with a peak intensity of

---

[1]A culvert is a drain or pipe that allows water to flow under a road or railway.

approximately 100 mm/h, whereas additional infrastructural improvements are needed to cope with larger rainfall events (exceeding 300 mm/day) (Dasgupta et al. 2015). The analyses of rainfall data for Dhaka have shown that the return period of extreme rainfall events is lessening, and 100-year rainfall events like the one in September 2004 (341 mm/day) will likely become more common in the future due to climate change. For example, heavy and persistent rainfall in June 2015 sub-merged parts of Dhaka for a long time and caused suffering for many city dwellers (The Daily Star 2015). Although the structural measures provided some protection from flood incidents, they faced many problems relating to their effectiveness and function during extreme rainfall events. This most recent example clearly indicates that we are not prepared for the unforeseen climatic events that might disrupt the whole city fabric.

11.5 Non-structural Measures

Non-structural measures should be an integral part of urban flood management along with the structural options to reduce damage and enhance effective response. In the case of Dhaka, it is evident that traditional structural measures are not sufficient for guaranteed flood protection (Faisal et al. 1999). Over time, the focus has shifted towards non-structural options. The following section will highlight some of the non-structural methods implemented in Dhaka. Many of these measures were proposed after the catastrophic flood in 1998 (Faisal et al. 1999).

11.5.1 Creation of Flood Zones and Building Regulations

According to the definition of the Bangladesh National Building Code (BNBC) standards, if a zone has the possibility of getting flooded with a water level of one meter or more, it is considered a flood-prone area. The code recommended that any building constructed in those areas should have the lowest floor above the flood zone (Mozumdar 2005). The Detail Area Plan (DAP) of the Dhaka Metropolitan Development Plan (DMDP) and the Urban Area Plan (UAP) of the city also have precise demarcation of flood flow areas. The flood flow zones demarcated by the DAP restrict possibilities of any future large-scale development (Morshed 2013).

The zoning of flood-prone areas and land-use restriction implemented by RAJUK through the DAP should work as an effective measure towards keeping people away from flood incidents. However, much of the low-lying areas of the city (considered as wetlands) are encroached upon due to the overwhelming population needs (Braun/Aßheuer 2011). A significant portion of these developments is being constructed by housing developer agencies, which violates many regulations. However, through financial and political backup these agencies are building housing that will effectively nullify the implication of DAP as a flood mitigation policy (Morshed 2013). In addition to the construction made by land developers, illegal slums are being built which are at high risk from flooding due to their exposure and inherent vulnerability (poorer residents are vulnerable due to lack of stable jobs and financial crisis). An active zoning policy might include restricting these illegal settlements and relocating people into flood free areas. Box 11.1 outlines flood vulnerability of slum dwellers, their preferences, and relocation options.

Box 11.1: Slum Dwellers' Relocation Preference. *Source*: The authors.

This study conducted by Rashid et al. (2007) aimed to formulate a preference model chosen by slum respondents if they were asked to relocate to flood free zones. The central hypothesis of the study was that if the slum dwellers in case study areas were given economic incentives, they would relocate to flood free areas.

The study was based on an in-depth questionnaire survey conducted in two areas in Dhaka city: Mirpur and Vasantek . The survey respondents were selected through a systematic approach, with 200 respondents from each of the slums.

The findings of the survey were grouped into three categories. The first is how the respondents assess the flood problem. The result shows that most people have faced some level of damage during a flood incident, and the situation is worse in the Vasantek area. The second is the socio-economic conditions of the slum dwellers, many of whom migrated from rural areas, and work as factory laborers or rickshaw pullers. In both locations, they faced infrastructural and political impediments. The third finding is that the population of these localities would only relocate from the current position to a flood free zone if they were given several economic incentives such as flood-free land at no cost, employment opportunities, grants, and loans to relocate to the new site. If these incentives were not present, then the residents were not willing to move.

The study also revealed that the local people are strongly tied to their current location, despite harassment from government agencies and local extortionists. Therefore, the case study concluded that it would require significant economic and structural incentives to relocate people, and sometimes it might be more efficient to improve on-site coping strategies such as in-situ upgrading of slums through the provision of infrastructure, water supply and sanitation, and management of pollution in the slums.

11.5.2 Flood Forecasting and Warning

Flood forecasting and warning systems are efficient ways to reduce the damage resulting from flood incidents (Dasgupta et al. 2015). The Flood Forecast and Warning Center (FFWC) of BWDB provides live daily updates of river water levels at different points throughout the country and disseminates the information when it goes above the danger mark. In Dhaka, the forecast and warnings are mainly calculated and distributed for river water flooding.

Although the center generates regular forecasts, the technical expertise remains limited and the information and warning messages are confusing for local people to understand. The messages from the monitoring station are primarily designed for decision makers and flood experts, making it difficult for non-technical or illiterate persons (especially residents in the slum areas) to respond. Also, with a lack of understanding of the potential impacts of different water depths, people find it difficult to take protective measures. The warning information only provides the water depth of the rivers but they do not give any spatial reference; thus people find it hard to adapt.

11.5.3 Flood Shelter

Dhaka experiences inundation each year due to river flooding in the eastern part of the city and water logging caused by drainage congestion in the west (Bala et al. 2009). This constant inundation is creating the need for flood shelters to provide temporary accommodation for the affected people. Most of the shelters in Dhaka are built on an ad hoc basis after the flood incidents (Maniruzzaman/Alam 2002) and are not planned and duly designated or maintained like the rural ones; this might create severe problems in emergency evacuation and management during sudden and long-term flood incidents (Masuya et al. 2015). Information about the limited number of designated shelters is also not well disseminated, thus crippling the emergency operation. In a recent study, Masuya et al. (2015) conducted a spatial analysis of plausible flood shelters in the city and their proximity to vulnerable residential areas (summarized in Box 11.2). The result verifies the importance of the flood shelters in Dhaka.

> **Box 11.2:** Potential Location of Flood Shelters in Dhaka City
>
> The case study was designed to achieve three objectives: to show the spatial distribution of potential flood shelters in the city of Dhaka, to identify vulnerable housing units, and to assess the potential of these flood shelters in terms of their ability to save the people during disasters. The study was conducted by Masuya et al. (2015) in a subset of the Dhaka Metropolitan Development Plan (DMDP) zone.

First, the survey team developed a raster map of the case study area and assigned a flood rank for each of the cells, thus creating a flood hazard map for the area. The rank was allocated according to flood frequency and flood depths. The team then identified the potential flood shelters and vulnerable households. The flood shelters were selected by building type, number of stories, and distance from the flood hazard. On the other hand, the vulnerable families were chosen by the structure type of their home and their location with respect to flood hazard intensity. Finally, the study calculated the average number of people to be saved by the potential flood shelters.

The result of the analysis is described in three segments. According to the spatial distribution of the flood hazards in the city, the eastern side is located in the high hazard zone. The analysis also reveals that around 60% of the study areas fall under hazard zones, with 45% classed as highly hazardous. The second part of the analysis identified 5537 buildings that can be used as flood shelters of different sizes. The distribution of these shelters was shown (Masuya et al. 2015) in conjunction with vulnerable households of five catchment areas, indicating how many households have access to the shelters. The results suggest that none of the catchment areas have enough shelters for the overall vulnerable population.

The implications of the study are significant as they can help improve the emergency evacuation plan for Dhaka city during a flood. However, more research is needed to assess the condition and capacity of the potential shelters.

11.6 Conclusion

Flood risk implications should be embedded into city planning processes as flood damage costs are significantly high. As discussed, flooding poses a significant threat towards Dhaka, and the city lacks the proper mechanisms to counter it. The structural flood protection measures are in a poor condition. The embankment to the west of the city lacks adequate and continuous maintenance. The eastern part of the city still does not have any physical protection from flooding. The city also lacks the ability to counter illegal encroachment and land acquisition around the dam and low-lying areas, which reduce natural water drainage. The housing and commercial developments do not follow proper regulation, and there is a lack of adequate communication regarding flood risk and the need for well managed and equipped flood shelters. All these factors increase the vulnerability of the city and decrease the ability to cope with extreme flood events.

The large-scale climatic and economic migration from all over Bangladesh creates significant pressure to land availability in the city and forces people to live in the low-lying, flood-prone areas. Many of these settlements also do not have any legal rights, and thus these groups need special attention in the overall flood management scenario. In short, the flood risk in Dhaka is a combined result of its climatic and geographical conditions and its weak political and economic capacity to address the situation.

As a result, improvements in both structural and non-structural measures are needed to counter the risks posed by flooding in Dhaka. The eastern part of the city has to come under the protection of a flood embankment. In addition, proper maintenance and monitoring of the embankments must be ensured. Having a long-term financial mechanism in place is also a critical factor. As several organizations are involved in the overall management process, perhaps an umbrella organization mainly designed for flood risk management can be created to ensure multi-stakeholder coordination. Stronger legal enforcement and political willingness are required to recover illegally occupied canals and river banks.

An effective early warning system is needed to ensure that people have enough time to prepare for any upcoming flood events. The warning messages should be clear and disseminated properly. Each ward and community in the city should have a flood management committee to help local people understand the implications of such disasters. Flood shelters should be created throughout the city, particularly close to the flood-prone areas where most of the migrated, low-income people live. These shelters should be managed well with the participation of each community and ward.

Finally, Bangladesh has several national level strategies, such as the National Plan for Disaster Management, National Adaptation Programme of Action (NAPA) and Bangladesh Climate Change Strategy and Action Plan (BCCSAP) to address the climate change and disaster situation. However, these policies lack specific guidelines for Dhaka and other urban areas. A tailored management system based on the national strategies is needed to achieve effective flood risk mitigation (in Dhaka), which will include the reduction of hazard exposure and improvement of the overall coping capacity of vulnerable communities.

References

Ahmed, B., Kamruzzaman, M., Zhu, X., Rahman, M. S. and Choi, K. (2013). Simulating Land Cover Changes and Their Impacts on Land Surface Temperature in Dhaka, Bangladesh. *Remote Sensing*. 2013; 5(11):5969–5998 .

Alam, M. and Rabbani, M. G. (2007). Vulnerabilities and responses to climate change for Dhaka. *Environment and Urbanization* 19(1):81–97.

Andjelkovic, I. (2001). Guidelines on non-structural measures in urban flood management. In *Technical documents in hydrology* (No. 50). UNESCO, Paris.

Bala, S. K., Islam, S., Chowdhury, J., Salehin, M. (2009). Performance of flood control works around Dhaka city during major floods in Bangladesh. Paper presented at the 2nd international conference on water and flood management, Dhaka, 15–17 March 2009.

Braun, B. and Aßheuer, T. (2011). Floods in megacity environments: vulnerability and coping strategies of slum dwellers in Dhaka/Bangladesh. *Natural Hazards* 58:771–787.

Cities Alliance (2016). Climate migration drives slum growth in Dhaka; at: http://www. citiesalliance.org/node/420 (24 June 2016).

Chowdhury, M. R. (2003). The Impact of 'Greater Dhaka Flood Protection Project' (GDFPP) on Local Living Environment–The Attitude of the Floodplain Residents. *Natural Hazards* 29(3): 309–324.

Dasgupta, S., Zaman, A., Roy, S., Huq, M., Jahan, S. and Nishat, A. (2015). *Urban Flooding of Greater Dhaka in a Changing Climate: Building local resilience to disaster risk*. World Bank, Washington.

Dewan, A. (2010). Bloated Dhaka (2010). In: *The Daily Star*, 18 February.

Dewan, A. M. (2013). *Floods in a megacity: geospatial techniques in assessing hazards, risk and vulnerability*. Dordrecht: Springer.

Dewan, A. M. and Corner, R. J. (2012). The impact of land use and land cover changes on land surface temperature in a rapidly urbanizing megacity. In: *IEEE International Geoscience and Remote Sensing Symposium* (IGARSS), Munich, 22–27 July 2012.

Faisal, I. M., Kabir, M. R. and Nishat, A. (1999). Non-structural flood mitigation measures for Dhaka City. *Urban Water* 1(2): 145–153.

Gain, A. K., Mojtahed, V., Biscaro, C., Balbi, S. and Giupponi, C. (2015). An integrated approach of flood risk assessment in the eastern part of Dhaka City. *Natural Hazards* 79:1499–1530.

Griffiths, P., Hostert, P., Gruebner, O. and Van der Linden, S. (2010). Mapping megacity growth with multisensor data. *Remote Sens Environ* 114(12):426–439.

Haque, A. N., Grafakos, S. and Huijsman, M. (2010). *Assessment of adaptation measures against flooding in the city of Dhaka, Bangladesh*. In IHS Working Papers (No.IHS WP 25), Rotterdam, Netherlands.

Huq, S. and Alam, M. (2003). Flood management and vulnerability of Dhaka City. In Kreimer, A., Arnold, M. and Carlin, A. (eds) *Building Safer Cities: The Future of Disaster Risk*. Washington, DC, pp. 121–135.

IPCC (2014). Emergent risks and key vulnerabilities. In: *Climate Change 2014: Impacts, Adaptation, and Vulnerability. Part A: Global and Sectoral Aspects. Contribution of Working Group II to the Fifth Assessment Report of the Intergovernmental Panel on Climate Change*. Cambridge University Press, Cambridge, United Kingdom and New York, NY, USA, pp. 1039–1099.

Kabir, A. and Parolin, B. (2012). Planning and development of Dhaka–a story of 400 years. In *15th International Planning History Society Conference. Cities, Nations and Regions in Planning History*, Sao Paulo (pp. 1–20).

Mahmud, M. S., Masrur, A., Ishtiaque, A., Haider, F. and Habiba, U. (2011). *Remote Sensing & GIS Based Spatio-Temporal Change Analysis of Wetland in Dhaka City*, Bangladesh. Journal of Water Resource and Protection 3(11).

Maniruzzaman, K. M. and Alam, B. M. (2002). A study on the disaster response for shelters during the 1998 flood in Dhaka city. *Engineering concerns of flood*, pp. 187–200.

Masuya, A., Dewan, A. and Corner, R. J. (2015). Population evacuation: evaluating spatial distribution of flood shelters and vulnerable residential units in Dhaka with geographic information systems. *Natural Hazards* 78:1859–1882.

Mojtahed, V., Giupponi, C., Biscaro, C., Gain, A. K. and Balbi, S. (2013). *Integrated assessment of natural hazards and climate-change adaptation: II. The SERRA methodology*. Working papers no. 07/WP/2013, Department of Economics, Ca' Foscari University of Venice.

Morshed, M. M. (2013). Detailed Area Plan (DAP): Why It Does Not Work? Planned Decentralization: Aspired Development: Souvenir published on World Habitat Day. Available at: http://www.bip.org.bd/SharingFiles/journal_book/20140128161651.pdf.

Mozumder, P. (2005). *Exploring Flood Mitigation Strategies in Bangladesh*. Professional Project Report, University of New Mexico.

Murshed, S. B., Islam, A. K. M. S. and Khan, M. S. A. (2011). Impact of climate change on rainfall intensity in Bangladesh. In: *Proceedings of the 3rd International Conference on Water and Flood Management, Dhaka, Bangladesh, 8–10 January 2011*.

Rashid, H., Hunt, L. M. and Haider, W. (2007.) Urban flood problems in Dhaka, Bangladesh: slum residents' choices for relocation to flood-free areas. *Environmental Management*, 40(1): 95–104.
The Daily Star (2015). Incessant rainfall inundates city roads; at: http://www.thedailystar.net/incessant-rainfall-inundates-city-roads-37116 (20 June 2016).
UNISDR (2015). Terminology; at: http://www.unisdr.org/we/inform/terminology (10 Sept 2015).

Chapter 12
Solar Home Systems in Bangladesh

Maliha Muzammil and Raihan Uddin Ahmed

Abstract Access to electricity is an important part of the Sustainable Energy for All (SE4ALL) initiative under the United Nations. Globally about 1.2 billion of people are deprived of electricity, mostly concentrated in rural areas of developing countries including Bangladesh. About two-thirds of the rural population in Bangladesh are yet to enjoy this basic utility. Grid-based electricity is challenging for various reasons including technical, and commercial issues. Development partners including the World Bank started a formal approach for promoting renewable energy and more than 3 million solar home systems have been installed in the off-grid areas of Bangladesh. By replacing kerosene, solar energy has not only facilitated people with an option of a much greater quantity of far-higher quality of lighting, at a lower cost, it has reduced the safety and health risks associated with kerosene, particularly among women and young children. However, considering that only about 10% of people in off-grid areas have adopted solar systems to date, the remaining potential for carbon emission reduction and adaptation is large. In Bangladesh, solar systems are a proven renewable energy technology in dealing with climate change and enhancing socio-economic development. However, there are some remaining technical, commercial and attitudinal constraints which need to be addressed to ensure sustainable development.

Keywords Solar home systems · Electricity · Rural · Energy · Kerosene Solar

Maliha Muzammil, Environmental Change Institute, University of Oxford, South Parks Road, Oxford, OX1 3QY, United Kingdom, Corresponding Author, e-mail: maliha.muzammil@ouce. ox.ac.uk.

Raihan Uddin Ahmed, Infrastructure Development Company Limited (IDCOL), UTC Building, 16th Floor, 8 Panthapath, Kawran Bazar, Dhaka-1215, Bangladesh.

© Springer Nature Switzerland AG 2019
S. Huq et al. (eds.), *Confronting Climate Change in Bangladesh*,
The Anthropocene: Politik—Economics—Society—Science 28,
https://doi.org/10.1007/978-3-030-05237-9_12

12.1   Introduction

Bangladesh has made remarkable progress in the installation of solar home systems (SHS) in rural areas. Bangladesh has been able to create 114,000 jobs, making it the 6th largest renewable energy-related workforce in the world in 2013 (IRENA 2014), similar in size to Spain's, according to a recent report by the International Renewable Energy Agency. According to Christine E Kimes, acting head of the World Bank in Bangladesh in 2014, the SHS programme has led to a reduction of carbon dioxide emissions by more than 538,000 tonnes a year. This chapter will set out the historical background to solar energy and then go on to analyse financial and institutional aspects and ambient regulatory measures. In addition, this chapter will try to reflect on the benefits of SHS and the low carbon, resilient development (LCRD) opportunities that they bring for Bangladesh, as well as the issues that are appearing as emerging challenges.

In Bangladesh, only 62% of the population have access to grid electricity, and generation per capita is one of the lowest at 321 kWh per annum (Islam 2014). Almost 15 million rural households use kerosene lamps in their homes in the absence of electricity (Rai et al. 2015b). To overcome these challenges, the *Solar Home System* (SHS) programme in Bangladesh has grown to be one of the largest off-grid electrification initiatives in the world (Khandker et al. 2014). Moreover, with more than 70,000 solar home systems being installed every month, the programme has been described by the World Bank as the 'fastest growing SHS programme in the world' (World Bank 2014). The Infrastructure Development Company Ltd. (IDCOL) started its SHS programme in 2003 to ensure access to clean electricity for the energy starved rural areas of Bangladesh (IDCOL 2014). As of April 2014, three million units had been installed under IDCOL's SHS programme (Khandker et al. 2014; IDCOL 2014). IDCOL now has a new target of reaching 6 million SHS beneficiaries by 2017, with an estimated capacity of 220 MW (IDCOL 2014). The fact that IDCOL outgrew its own target of 50,000 units installed in 5 years, within a month and went on to install three million units by 2014 (Khandker et al. 2014) attracted both government and international donor funding in the long term.

IDCOL's success in ensuring such large-scale coverage was possible due to the subsidies provided by donors to facilitate SHS adoption in remote and off-grid areas (Khandker et al. 2014). Around 10% of off-grid areas have been reached, which means there is still ample scope for SHS expansion (Khandker et al. 2014). IDCOL's SHS programme has been particularly successful because it combines price support with quality assurance, installation, and after-sales support as a one stop solution to households (Khandker et al. 2014).

12.2   Basics of Solar Home System Technology

A solar home system (SHS) offers households in developing countries such as Bangladesh a convenient supply of electricity for lighting and running small appliances (e.g., small television sets, radios, and mobile phone chargers) for about 3–5 hours a day, using energy from sunlight. Typically, an SHS consists of a small solar photovoltaic (PV) panel, charge controller, battery, compact fluorescent lamp (CFL) or light-emitting diode (LED) lights, and a universal outlet for charging mobile phones and small appliances (Fig. 12.1) (Khandker et al. 2014).

The solar panel, also known as the photovoltaic (PV) module, is the heart of any SHS. Usually installed on the roof of a house at an angle designed to collect maximum sunlight, it converts sunlight into electrical energy. The rechargeable battery stores electricity for use at night and on cloudy days, and provides the voltage needed to run appliances; in Bangladesh, appliances are designed for 12 volt (V) operation (Khandker et al. 2014). The charge controller, positioned between the solar panel and the battery, protects the battery against overcharging (e.g., on bright sunny days) and discharging below a certain cut-off voltage, which can cause permanent damage. Watt-peak (Wp) is the unit of measure used to express the capacity or power generated by the SHS. The capacity range for most SHS units installed in Bangladesh is 20–120 Wp. A system with a 50 Wp capacity can power four lights, a mobile phone charger, and a television set (Khandker et al. 2014).

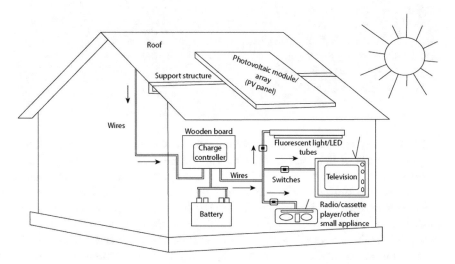

Fig. 12.1 Schematic diagram of an operational SHS. *Source* Khandker et al. (2014: 11)

12.3 History and Background of Solar Energy in Bangladesh

In the 1980s, an assessment about the application and technological aspects of *solar home systems* (SHS) was initiated at *Bangladesh University of Engineering and Technology* (BUET), *Dhaka University* (DU) and *Bangladesh Atomic Energy Commission* (BAEC). In 1988 BAEC launched the Sandwip Solar PV Demonstration Program with money from the Bangladesh Government, which operated satisfactorily until it was significantly damaged by cyclones and water surges in April 1991. In 1993, the *Rural Electrification Board* (REB) was initiated to promote Renewable Energy Technologies in the Karimpur and Nazarpur unions under the Narsingdi Sadar Thana project of Narsingdi District, with financial support from the French Government (Islam 2012). The *Local Government Engineering Department* (LGED) subsequently installed solar systems in 1998 with financial support from UNDP. The *Infrastructure Development Company Ltd.* (IDCOL) also started to promote dissemination of SHS in remote rural areas through the Rural Electrification and Renewable Energy Development Project with financial support from the World Bank, *Global Environment Facility* (GEF), KfW, GTZ, ADB and IDB from 2003 (Islam 2012).

Box 12.1: IDCOL SHS at a Glance. *Source* **IDCOL (2016).**

Programme started: January 2003
Programme target: 6 million SHSs by 2018
Programme achievement: 3.6 million by April 15
No. of beneficiaries: 15.50 million people
Fossil fuel Saving: 200,000 ton/year worth $ 200 million
Job creation: 70,000 people
IDCOL investment: USD 600 million

Solar energy is assumed to have the potential to serve a central role in solving the power crisis in Bangladesh. Bangladesh is situated between 20.30 and 26.38 degrees north latitude and 88.04 and 92.44 degrees east, which is an ideal location for solar energy utilisation (Hoque et al. 2013). Bangladesh receives an average daily solar radiation of 3.82–6.42 kWh/m^2 (Ullah et al. 2012). The country has a total area of 147,570 km^2. With a solar system efficiency of 10%, a total of 5.2×109 kWh units of electricity can be generated annually (Khan/Khan 2009).

Ensuring access to electricity at a satisfactorily level is a pivotal challenge for Bangladesh in achieving its vision of progressing from its recent status of becoming a lower middle income country to becoming a middle-income country by 2021 (World Bank 2015), yet nearly three-fifths of rural households still lack access to electric power. The country's limited ability to generate and distribute enough grid-based electricity to meet growing demand (Barnes 2007; Zerriffi 2011), combined with its ample sunshine and high levels of energy poverty, point to a large potential market for SHS in poorer off-grid areas (Khandker et al. 2014).

12.4   Financial and Regulatory Aspects of SHS in Bangladesh

Solar systems of 40–85 Wp in size are mostly used in the rural areas in Bangladesh. The cost of a 40 Wp system is around 17,000 Bangladeshi Taka (BDT), whereas that for an 85 Wp system costs about BDT28,000 in cash (IDCOL 2016). The average payback period is assumed to be 4.2 years and varies between 3.1 and 6.5 years (Hoque et al. 2013). The cash sale option is considered to be the best option to achieve the facility at the lowest cost. However, there are some other types of financing schemes that are provided by Partner Organisations to the customers, including the following (IDCOL 2016):

Option 1: 15% down payment and the remaining 85% payable in 36 monthly instalments with a flat rate service charge of 12%.
Option 2: 25% down payment and the remaining 75% payable in 24 monthly instalments with a flat rate service charge of 10%.
Option 3: 35% down payment and the remaining 65% payable in 12 monthly instalments with a flat rate service charge of 9%.
Option 4: 25% down payment and the remaining 75% payable in 12 monthly instalments without any service charges (only for mosques/temples/churches) (Fig. 12.2).

Component	% of total SHS system cost (%)
Battery	30.00
Solar PV panel	28.00
3 years after sales service and instalment collection	13.50
Overhead cost	10.00
Other accessories	7.50
Lampshades	5.00
Charge controller	3.00
Fluorescent lamps	2.00
Structure to install solar PV	1.00

SHS beneficiaries are required to make a down payment of only 10% to establish ownership of the system, while monthly instalments are made through a micro finance institution (MFI) to pay off the remainder. These monthly instalments are kept within the households' affordability range (Rai et al. 2015a). Previously, many

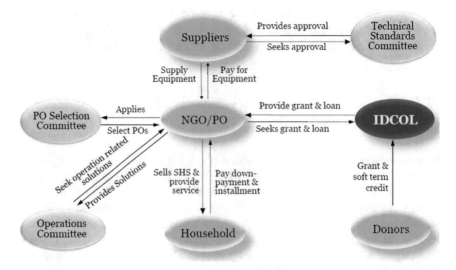

Fig. 12.2 Overview of SHS financing process of IDCOL. *Source* IDCOL (2016: page). Permission may be needed

banks had been unwilling to lend to the poor or required a large down payment with exorbitant interest rates. However, with IDCOL's support, the MFIs are able to access loans and provide the households with affordable credit (Rai et al. 2015a; Khandker et al. 2014). IDCOL also provides a refinance credit scheme so that MFIs can pay suppliers immediately. MFIs retain a 10% equity stake in the project, with the households owning another 10% equity and IDCOL providing the remaining 80% in the form of refinancing credit (Rai et al. 2015a). However, even though the upfront grant portion helps the poor gain access to solar energy, the most marginalised populations are still not able to afford it, as it is not specifically targeted to the ultra-poor populations (Rai et al. 2015b).

Some aspects of the Bangladesh SHS programme are quite unique and may be difficult to replicate in other countries (Sadeque et al. 2014): the SHS programme in Bangladesh benefitted largely from a strong pre-existing network of competitive microfinance institutions (MFIs) with a deep and wide-ranging reach in rural areas, and a well-developed system for rural households to access credit. Other factors contributing to the success of the programme include the high density of Bangladesh's rural population which allows for competition and economies of scale; the rising rural incomes and remittances from abroad; and the existence of entities interested in doing business with rural customers and the country's entrepreneurial culture (Sadeque et al. 2014).

12.4.1 Incentives that Enable or Constrain Low Carbon Resilient Development (LCRD) Investments

Low-carbon development is an approach that focuses on reducing greenhouse gas emissions through the development process. Resilience refers to building the capacity of society to recover after climate-related shocks and is associated with adaptation to climate change. As a result, low carbon resilient development involves bringing together the three policy areas of climate change mitigation, adaptation and development to find synergies and 'win-wins' at the level of a policy, an objective or within a financing mechanism (Fisher 2013). Low-carbon resilient development therefore seeks to link all three of these policy objectives in the context of national development (Fisher 2013).

The *Solar Home System* (SHS) programme provides low-carbon energy in supporting co-benefits between some of these agendas, even though it was not originally intended as a comprehensive low-carbon resilient development approach. *Bangladesh's Climate Change Strategy and Action Plan* (BCCSAP) includes both adaptation and mitigation approaches, and a plethora of strategic documents at the national level indicate a widespread interest by the government on how they can operationalise both aspects of the climate change agenda at the national level (Fisher 2013). Bangladesh's *Intended Nationally Determined Contributions* (INDCs) set out a number of mitigation actions that will help limit the country's GHG emissions. These mitigation actions will play a key role in realising the move to a low-carbon, climate-resilient economy and to becoming a middle-income country by 2021. Bangladesh's existing strategies and plans, in particular the Bangladesh Climate Change Strategy and Action Plan (BCCSAP), Renewable Energy Policy 2008, the Energy Efficiency and Conservation Master Plan (EE&C Master Plan), the forthcoming National Adaptation Plan, the National Sustainable Development Strategy, the Perspective Plan (Vision 2021) and the Sixth (and forthcoming seventh) Five Year Plan, the National Disaster Management Plan and the Disaster Management Act all lie at the heart of the INDCs (MoEF 2015). LCRD investments involve a wide range of actors in both policymaking and implementation. Delivery of investments is shaped by how these various actors work with ideas, power and resources to make and implement decisions.

Different actors have wide ranging incentives to invest in renewable energy in Bangladesh. The SHS programme primarily set out to achieve the objective of energy access. The government's policies and mandates acted as drivers for more widespread dissemination of SHS (Rai et al. 2015a). The SHS programme is backed by Bangladesh's national strategy, which calls for achieving universal access to electricity by 2021. Electricity has also been a critical input towards achieving the Millennium Development Goals (MDGs), affording households benefits such as clean energy for high-quality lighting, which improves health and enables children to study for longer periods after sunset, greater farm- and non-farm productivity, and women's empowerment through better time allocation and access to information (Khandker et al. 2014). Affordable and clean energy are also part of the

Sustainable Development Goals (SDGs) which aim to end poverty, fight inequality and injustice, and tackle climate change by 2030 (UNDP 2016).

The Sustainable and Renewable Energy Development Authority (SREDA) was also set up to further promote and develop the sustainable and renewable energy landscape. Fiscal incentives, such as reduced import tariffs and taxes on renewable energy products, and policies to encourage private sector investment in the power sector using the independent power producer (IPP) model (Rai et al. 2015a), were all important factors in enabling widespread growth of the SHS programme.

The Bangladesh Bank (BB) is the first central bank in the world to actively provide dedicated resources for sustainable development. The BB first set up a refinancing scheme for commercial banks on finance for green energy, including solar and biogas projects in 2005. In 2010, it introduced a US$26 million refinancing facility for investments in green energy and effluent treatment plants, thus allowing commercial banks to access capital at lower rates, increasing the profitability of green lending (Rai et al. 2015a). The Central Bank of Bangladesh allocates funds to commercial banks based on three mechanisms: refinancing, spontaneous financing and incentive-based financing (Masukujjaman/Aktar 2013 in Rai et al. 2015a).

12.5 Institutional and Policy Arrangements by the Bangladesh Government

The Renewable Energy Policy was approved by the Bangladesh government in December 2008 and became effective from 2009. The objectives of this policy are to harness the potential of renewable energy resources and disseminate it to the people, and to enable, encourage and facilitate both public and private sector investment. The policy has set the target of generating 5% of electricity (800 MW) by end of 2015 and 10% of electricity by end of 2030 from renewable energy sources (GoB 2008; Power Division 2013; Christian Aid 2014). Solar energy is expected to contribute to around 500 MW of renewable electricity in order to achieve the 800 MW target (GoB 2008). Key objectives of the Renewable Energy Policy (GoB 2008) include: harnessing the potential of renewable energy resources and dissemination of renewable energy technologies in rural, peri-urban and urban areas; facilitating private sector investment in renewable energy projects; and scaling up contributions of renewable energy to electricity production. In addition, the policy facilitates various types of tax and tariff waivers for the renewable energy industry in Bangladesh. In 2007, the government also approved the Remote Area Power Supply System (RAPSS) Guidelines for power generation and the distribution and supply of electricity in remote and isolated areas.

Bangladesh has developed a diverse set of policies to encourage energy access, the most recent of which is the government of Bangladesh's vision to ensure 'Electricity for all by 2021' (Power Division 2013). Up to 70% of Bangladesh's total commercial energy is provided by natural gas and the rest by imported oil

(Islam 2014). Natural gas is in short supply, which is another reason the government has been keen to push the renewable energy agenda forward. Access to electricity has also been a major factor achieving the Millennium Development Goals in Bangladesh (Khandker et al. 2014).

Recognising urgent energy challenges, Bangladesh joined the global coalition under the United Nations' Sustainable Energy for All Initiative, which calls on governments, businesses and civil society to achieve three goals by 2030 (World Bank 2012): universal access to energy, double the renewable energy share of power produced and consumed from 15 to 30%, and double the energy efficiency improvement rate.

Bangladesh also published an ambitious INDC in 2015, in which it has outlined its intention of an unconditional contribution to reduce GHG emissions by 5% *against Business as Usual* (BAU) levels by 2030 in the power, transport and industry sectors, based on existing resources (MoEF 2015). The INDC also includes a conditional 15% reduction in GHG emissions against BAU levels by 2030 in the power, transport, and industry sectors, subject to appropriate international support in the form of finance, investment, technology development and transfer, and capacity building (MoEF 2015).

In order to meet the unconditional contribution set above, Bangladesh already has a number of activities and targets driving action to reduce GHG emissions, including the Solar Homes Program. Under this programme, access to off-grid electricity is provided to rural areas and around 4 million Solar Home Systems have already been distributed across the country (MoEF 2015). In addition, aggressive targets have been set for scaling up the potential for solar irrigation pumps, solar mini and nano grids to address energy access in off-grid areas (MoEF 2015).

Keeping in line with the vision of ensuring universal access to electricity by 2021, addressing carbon emissions, as well as reducing diesel imports and subsidies and diversifying fuel options as outlined in the 2010 Power System Master Plan, the government of Bangladesh has been actively prioritising renewable energy by establishing a national policy and financial incentives to implement it (Rai et al. 2015a). The Bangladesh Bank, the Central Bank of the country, has a green energy portfolio, and IDCOL was set up to catalyse private sector renewable energy finance.

Bangladesh's government has created financial incentives for investment in the renewable energy sector, including 20-year tax holidays, reduced levies on importation of renewable energy technology and reduced taxes on local manufacture and assembly of renewable energy equipment (Islam 2014). Other incentives, including feed-in tariffs and incentives to attract foreign investment in the sector, are under consideration (Islam 2014). Moreover, the private sector has been allowed to generate electricity from renewable sources and sell to chosen customers at preferential rates (Islam 2014). Concessional finance and capital buy-down grants have also been made available for renewable energy projects (Rai et al. 2015a).

While implementing the Renewable Energy Policy 2008 and RAPSS Guidelines 2007, the need for an independent authority was widely felt. Accordingly, in December 2012, the government enacted legislation in Parliament to establish the SREDA. On 22 May 2014, SREDA was formally launched (SREDA 2014). The objectives of SREDA are to promote, develop and co-ordinate renewable energy and energy efficiency programmes in the country. This institution will also prepare short, mid and long term plans to meet government targets. It has set a target to achieve 2000 MW of electricity from renewable energy by 2021, i.e. 10% of total power generation. It also emphasises the need to improve energy intensity in 2030 by 20% compared to the 2013 levels, which is expected to result in savings of 95 million tonnes of energy during this period. Energy savings will total BDT768 billion in total or an annual average of BDT51 billion at the current weighted average natural gas price (SREDA and Power Division 2015). Renewable energy wings have also been set up at other organisations to assist in the implementation of renewable energy programmes and activities.

Key Agencies and Actors

The *Sustainable and Renewable Energy Development Authority* (SREDA) was set up as a focal point for the development and promotion of sustainable renewable energy for Bangladesh and is at the centre of Bangladesh's renewable energy landscape (Rai et al. 2015a). SREDA monitors entities that promote and finance energy projects, and supports public-private partnerships in renewable energy projects (Power Division 2013).

All activities relating to rural and renewable energy fall within the remit of the *Ministry of Power, Energy and Mineral Resources* (MPEMR). While the administrative oversight and support to SREDA is provided by the Ministry of Power, Energy and Mineral Resources, the Ministry of Finance manages the budget for SREDA, otherwise known as the 'Pool Fund', which is supported by international co-operation. The Ministry of Finance is also in charge of the renewable energy tax incentives and provides capacity building to financial institutions involved (Rai et al. 2015b).

The Central Bank of Bangladesh is a key financial intermediary in the renewable energy landscape (Rai et al. 2015a). It is the primary regulator of the country's monetary and credit system, and oversees all banking and non-banking financial institutions. It recently diversified into green lending, providing concessional finance to the financial sector in the form of green credit.

The *Infrastructure Development Company Ltd.* (IDCOL) is a non-banking financial institution which, since its inception, has played a major role in Bangladesh's renewable energy development (IDCOL 2014). It is hosted by the *Ministry of Finance* and governed by an independent board of directors from the Ministry of Finance, the Ministry of Information and Communication Technology, and the Ministry of Power, Energy and Mineral Resources (Rai et al. 2015a). By using donor funding from both domestic and international sources, IDCOL offers a range of measures including grants, subsidies, concessional loans and technical services for SHS (Fig. 12.3).

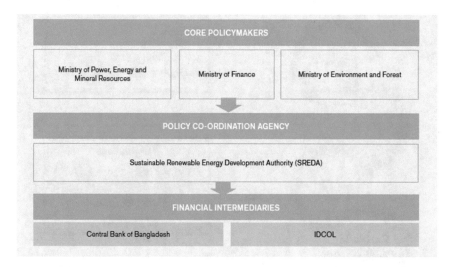

Fig. 12.3 Actors in Bangladesh's renewable energy policy landscape. *Source* Rai et al. (2015a: 15)

12.5.1 Institutional Mechanisms

IDCOL is a government-owned, public-private partnership, financial institution mandated to provide long-term financing for private infrastructure projects (Islam 2014; Rai et al. 2015a). It works in partnership with development partners, suppliers of solar home systems, SMEs, and participating MFIs, which are considered as Partner Organisations (PO). IDCOL's concessionary financing and grant support, technical assistance, quality assurance and capacity development of stakeholders were fundamental to the success of the SHS programme (Islam 2014).

IDCOL's innovative delivery model and the availability of and access to funding from donors were key in contributing to the major success of IDCOL's SHS Program (Rai et al. 2015a). Poor households are able to afford and access energy services by using microcredit financing from MFIs (microfinance institutions), as there are no upfront costs or payments for the operation and maintenance of the solar home systems (Rai et al. 2015a). IDCOL has used multiple means to target economic and financial barriers (as well as institutional, regulatory or information barriers), including grants, debts (concessional and non-concessional loans), equity and risk-mitigation instruments (Kato et al. 2014).

The SHS programme has been able to unlock funding in the form of long-term soft loans and equity (Rai et al. 2015a). Awareness of demand for SHS in off-grid areas, and the Bangladesh government's vision for energy access, have been key factors in supporting the programme's exponential expansion (Fig. 12.4).

The World Bank, *Asian Development Bank* (ADB), *Islamic Development Bank* (IDB), Department for International Development (DFID), Japan International Cooperation Agency (JICA), KfW Group, *Gesellschaft für Internationale*

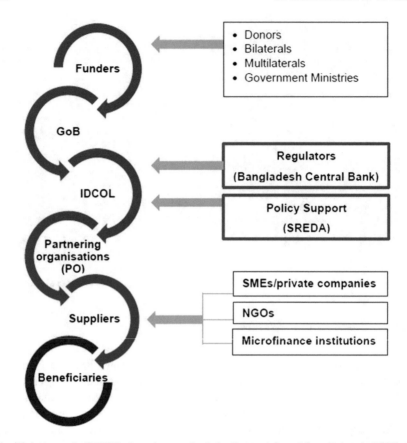

Fig. 12.4 Actors in IDCOL's financing supply chain. *Source* Adapted from Rai et al. (2015b: 34)

Zusammenarbeit (GIZ), *Global Environmental Facility* (GEF), *United States Agency for International Development* (USAID), and *Global Partnership on Output Based Aid* (GPOBA) all provide funding to IDCOL for the SHS programme, while the *Bangladesh Climate Change Resilience Fund* (BCCRF) provides its own grants (IDCOL 2014; Rai et al. 2014b).

IDCOL has a total of 47 Partner Organisations (POs) all over the country, which are responsible for selling and installing solar home systems. These include private SMEs, MFIs, and NGOs. Pos, which were initially chosen for their widespread coverage and offices in rural areas, along with their experience in microcredit programmes, have enabled far reaching energy access for the poor (Rai et al. 2015b). Partnering with organisations with better coverage in rural areas can help ensure credit disbursement, credit collection, and after sales services (Khandker et al. 2014). IDCOL recruits the POs, which are responsible for selecting potential SHS and Solar Irrigation Pump (SIP) buyers in off-grid areas, installing the systems, providing after sales service and maintenance, and developing a robust market chain (Khandker et al. 2014).

Another reason for SHS programme success is the stringent screening by IDCOL's PO selection committee, which assess the POs against eligibility criteria for inclusion in the programme. IDCOL has a technical standards committee that approves the suppliers and the SHS equipment to be used. The SHS programme has made systems affordable through a combination of consumer credit and (declining) subsidies (Khandker et al. 2014).

In order to keep the system process affordable, IDCOL provides POs with capital buy-down grants; through market competition, the grants are passed on to household buyers in the form of a lower price. Buyers are also offered microcredit to make SHS more affordable. All these incentives work together to create a market chain that ensures quality products that are affordable and locally serviceable (Khandker et al. 2014).

12.6 Benefits and Growth of Solar in Bangladesh

The SHS programme has been particularly successful because it combines price support with quality assurance, installation, and after-sales support as a one stop solution for households (Khandker et al. 2014). IDCOL involves multiple stakeholders, uses multiple instruments, and targets multiple barriers to increased deployment of low carbon and resilient SHS (Kato et al. 2014).

Over the years, core agencies such as the Central Bank of Bangladesh have also begun to regulate and channel bank and non-banking finance towards renewable energy investments (Rai et al. 2015b). The bank plays both a regulatory and a licensing role with all financial intermediaries, commercial banks, and financial institutions within the country, including the regulation of IDCOL (Rai et al. 2015a). Actions and policies implemented by the government of Bangladesh and the Central Bank of Bangladesh also align with the key policy areas for energy access.

Installing a SHS on the rooftop of a house can have immediate impacts: it enables the household to have light after nightfall, making it easier to study in the evenings, and allows people to watch TV and be informed of many useful and socially desirable things happening around them, perhaps inspiring them to take part in such activities (Khandker et al. 2014). Furthermore, it can lower levels of household air pollution (HAP) through reduced use of kerosene Solar electricity also has the potential positive externality of replacing fossil fuels for electricity generation, thus contributing to lowering carbon dioxide (CO_2) emissions and the harmful effects of climate change.

According to the World Bank (2012), by late 2012, an average 8% of off-grid rural households had adopted SHS. The divisions of Barisal and Sylhet had the highest take-up of SHS, at 13.4% and 13.2%, respectively; while the lowest adoption rates were found in Rajshahi and Rangpur, at 3.9% and 3.3%, respectively. Figure 12.5 clearly reflects the surge in SHS adoption in recent years, particularly from 2009 onwards. As of 2014, the country's capacity to generate

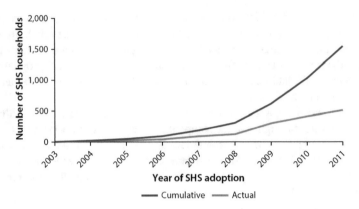

Fig. 12.5 SHS adoption by year from sample data. *Source* Khandker et al. (2014: 29)

electricity from renewable sources such as solar energy, biomass and wind-based plants, stands at 363.8 megawatts, which is 3.51% of the total power generation capacity (Ministry of Finance 2015).

In its inception year (2003), IDCOL installed 11,697 SHS, and between then and 2006, there was linear growth in the adoption of SHS by rural people. This take-up of SHS grew exponentially between 2007 and 2013. However, as the technology and management are well structured, a steady trend in adoption is projected from 2014 to 2018 (on an average about 600,000 SHS annually), which is around 500% higher than was seen at inception.

Between 2012 and 16, the Bangladesh government has opted to generate 500 MW of energy from renewable energy sources; this is known as the 500 MW programme (SREDA 2016). Through this programme, 340 MW of energy is planned to be generated from commercial solar power projects, whereas, in the social sector, solar power projects are expected to contribute about 160 MW. SHS has been considered for commercial solar power projects. The other projects under this programme include solar irrigation, mini-grids and solar parks. Out of a total of 500 MW, the estimated contribution of SHS is about 30 MW (Islam 2012).

12.7 Emerging Challenges and Opportunities

The SHS technology is now familiar to about 12 million rural people in Bangladesh. To ensure quality control, a testing lab with international standards has been set up at BUET. Policies including the Renewable Energy Policy, SREDA, the 500 MW Goal, tax waivers for SHS accessories and soft loan facilities through Bangladesh Bank, demonstrate strong government commitment to the sustainability of SHS. The availability of a considerable number of competent solar accessory suppliers indicates the sustainability of the industry in Bangladesh. At present, energy utilisation in Bangladesh is about 0.15 W/m^2 land area whereas the availability of solar energy is above 208 W/m^2 (Assignment point 2015). Above all, there is still the bitter reality of the existence of about 50 million rural Bangladeshis without electricity, which clearly reflects the market demand (Assignment point 2015).

Currently, in addition to the IDCOL SHS programme, a number of other SHS initiatives (both formal and informal) are operating in Bangladesh, where reliability or sustainability have appeared as concerns. The introduction of low quality batteries, panels and accessories have become a concern. Apart from the technological challenges of marketing SHS in remote areas, financing a solar power programme is challenging for a number of reasons. Given the cost of units and distribution problems, many poor households would find it difficult to purchase a solar unit out of their own pockets (Shahidur et al. 2013). Above all, due attention is required for the proper disposal of expired lead acid batteries and PV panels. The potential environmental health and safety hazards caused by improper disposal of expired lead acid batteries and PV panels could tarnish the success of SHS and the renewable energy industry. One recent challenge has been the rise of grid-connected electrical connections through *Bangladesh Rural Electrification Board* (BREB) since the end of 2015, with monthly new connections in range of 300,000–500,000 per month against earlier monthly connections of 100,000 per month. This may slow down take-up of new SHS.

12.8   Conclusions

Bangladesh's success in the field of off-grid renewable energy, specifically with SHS, has become a global model. The government, private sector and public entities have made significant progress in developing a sustainable renewable energy industry including SHS. However, according to Schwan (2011), there are some issues relevant to the economic and financial aspects, market performance and legal and regulatory aspects, which require further attention to increase the effectiveness of SHS in Bangladesh. For example, subsidies for fossil fuels, limited access to credit and inadequate distribution networks, as well as lack of industry-wide standards (Schwan 2011). Further research would be needed to identify which of these barriers still remain an issue for further expansion of SHS in Bangladesh.

The IDCOL case study provides a number of key lessons that can be applied to energy access efforts in other countries (Rai et al. 2015b):

- *A strong policy foundation can drive the stakeholders along the value chain*: the uptake of SHS in Bangladesh was largely driven by strong policy measures, targets and fiscal measures from the national level, which created incentives for actors across all levels of the value chain. The recent creation of SREDA to promote the development and use of renewable energy is one such example of a strong push from the higher level towards low carbon, resilient renewable energy pathways.

- *An innovative, integrated, and holistic financing model can create win-win opportunities for all stakeholders in the value chain:* IDCOL's SHS programme offers a complete package that incentivises market creation, creating delivery networks, access to capital, quality assurance, after-sales service, training and institutional strengthening support for partner organisations.
- *Increased support from donors allowed the programme to reach the poor, enhance market development, and catalyse finance for smaller players:* Grants and subsidised credit were crucial within the IDCOL model to make SHS affordable for the target population and the poorer beneficiaries. Grants, low-interest loans, and microcredit arrangements also helped small-scale enterprises that have difficulty accessing finance from commercial markets access that finance.
- *IDCOL's success lies within its transparent and accountable system:* Its Technical Standards Committee, the partner organisation (PO) selection committee, and the monitoring and inspection team that reviews the reports from the PO, were essential in gaining increased credibility and support from the donors.
- *Enforcing technical quality:* Strict quality standards, a 20-year warranty for the solar panel and a 5-year warranty for batteries, and strong enforcement of these standards ensured the uptake and popularity of more SHS amongst consumers (Rai et al. 2015b; Sadeque et al. 2014).
- *Financing instruments used in a sequential model can help in developing a long term sustainable financing structure that remains viable* (Rai et al. 2015b): Due to the demand of SHS in Bangladesh a phase-out subsidy model and concessionary to semi-commercial credit have helped in the transition to a more sustainable financing arrangement, once the market has been developed.
- *Availability of a buy-back scheme:* The presence of a buy-back scheme, which ensures that the PO will repurchase the solar home system within a year if the household obtains a grid connection, allows for further expansion (Sadeque et al. 2014).

References

Amin, A. 2013. Long-term Finance: Enabling environments and policy frameworks related to climate finance.

Asaduzzaman, M., Yunus, M., Haque, A.K.E., Azad, A.K.M.A.M., Neelormi, S. and Hossain, M. A. 2013. *Power from the Sun: An Evaluation of Institutional Effectiveness and Impact of Solar Home Systems in Bangladesh.* Report submitted to the World Bank, Washington, DC.

Assignmentpoint.com 2015; at: http://www.assignmentpoint.com/science/importance-and-impact-of-solar-home-system-shs-in-bangladesh.html (6 September 2016).

Chairman's Message, 2014; at: http://www.sreda.gov.bd/index.php/chairman-s-message (6 September 2016).

Christensen, K., Raihan, S., Ahsan, R., Uddin, A.M.N., Ahmed, C.S. and Wright, H. 2012. *Financing Local Adaptation: Ensuring Access for the Climate Vulnerable in Bangladesh. Dhaka:* ActionAid Bangladesh, Action Research for Community Adaptation in Bangladesh, Bangladesh Centre for Advanced Studies, and International Centre for Climate Change and Development.

Fisher, S. 2013. *Low carbon resilient development in the least developed countries.* IIED Issue Paper. IIED, London.

Government of Bangladesh 2008. *Renewable Energy Policy of Bangladesh*; at: http://www. c4learn.com/current-affairs/2015/bangladesh-became-first-country-to-receive-funds-for-solar-home-systems-from-un.html (6 September 2016).

Hoque, S.M.N. and Das, B.K. 2013. Analysis of Cost, Energy and CO_2 Emission of Solar Home Systems in Bangladesh. *International Journal of Renewable Energy Research*, 3, No. 2.

IDCOL, 2016. Infrastructure Development Company Limited. Internal information.

Infrastructure Development Company Limited 2015; at: www.idcol.org.

Islam, S.M.F. 2014. "Financing Renewable Energy in Bangladesh" Presentation by Deputy CEO of Infrastructure Development Company Limited (IDCOL) at the West African Power Industry Convention, Lagos, November 2014.

Islam, Md.A. 2012. *Past Experiences and Current Challenges of Solar Energy Usage in Bangladesh*, Loughborough University, UK.

Kato, T., Ellis, J., Pauw, P. and Caruso, R. 2014. *Scaling up and replicating effective climate finance interventions.* Climate change expert group, Paper No. 2014 (1), OECD.

Fayyaz Khan, M., and Mahmud Khan, M. 2009. *Journal of Electrical Engineering*, The Institution of Engineers, Bangladesh, Vol. EE 36, No. II, December 2009.

Khandker, S.R., Samad, H.A., Sadeque, Z.K.M., Asaduzzaman, M. Yunus, M. and Haque, A.K.E. 2014. *Surge in Solar-Powered Homes: Experience in Off-Grid Rural Bangladesh. Directions in Development; Energy and Mining.* Washington, DC: World Bank Group; at: http://documents.worldbank.org/curated/en/2014/10/20286806/surge-solar-powered-homes-experience-off-grid-rural-bangladesh (6 September 2016).

Khandkere, S.R. et al. 2013. *The benefits of solar home system in Bangladesh-An Analysis from Bangladesh.* World Bank Group, Washington D.C, USA.

Khandkere, S.R., Khandker, S.R., Samad, H.A., Sadeque, Z.K.M., Asaduzzaman, M., Yunus, M. and Haque, M. 2014. *Surge in Solar-Powered Homes-Experience in Off-Grid Rural Bangladesh.* World Bank Group, Washington D.C, USA.

Ministry of Finance. 2015. *Bangladesh Economic Review 2014.*

MoEF. 2015. Intended Nationally Determined Contributions; at: http://www4.unfccc.int/submissions/INDC/Published%20Documents/Bangladesh/1/INDC_2015_of_Bangladesh.pdf (6 September 2016).

Rai, Neha, Asif Iqbal, Asif Iqbal, Antara Zareen, Tasfiq Mahmood, Maliha Muzammil, Saqib Huq, Noor Elahi. 2015a. *Financing Inclusive Low-Carbon Resilient Development: Role of Central Bank of Bangladesh and Infrastructure Development Company Limited*, Country Report. London: International Institute for Environment and Development; at: http://pubs.iied.org/10139IIED.html (6 September 2016).

Rai, Neha, Terri Walters, Sean Esterly, Sadie Cox, Maliha Muzammil, Tasfiq Mahmood, Nanki Kaur, Lidya Tesfaye, Simret Mamuye, James Knuckles, Ellen Morris, Merijn de Been, Dave Steinbach, Sunil Acharya, Raju Pandit Chhetri, and Ramesh Bhushal. 2015b. *Policies to Spur Energy Access: Volume 2: Case Studies of Public-Private Models to Finance Decentralized Electricity Access.* Golden, CO: National Renewable Energy Laboratory.

Sadeque, Z., Rysankova, D., Elahi, R. and Soni, R. 2014. Scaling up access to electricity: the case of Bangladesh. Live Wire 2014/21, World Bank Group, 2014; at: https://openknowledge. worldbank.org/bitstream/handle/10986/18679/887020BRI0Live00Box385194B00PUBLIC0. pdf?sequence=1 (6 September 2016).

Schwan, S. 2011. *Overcoming Barriers to Rural Electrification*, Masters Thesis, Aarhus University, Denmark.

SREDA. 2015. Sustainable and Renewable Energy Development Authority website; at: http://www. sreda.gov.bd/index.php/sustainable-energy/renewable-energy/500-mw-solar-development-program (6 September 2016).

SREDA and Power Division. 2015. *Energy Efficiency and Conservation Master Plan up to 2030.*

The World Bank, 2012. Sustainable energy for all: Gearing up for the road ahead; at: http://www.
    worldbank.org/content/dam/Worldbank/document/SDN/Sustainable_Energy_for_All_World_
    Bank_Outlook_2012.pdf (6 September 2016).
The Daily Star. 2015; at: http://www.thedailystar.net/frontpage/bangla-lesson-carbon-polluters-
    129517 August 20, 2015 (6 September 2016).
The Daily Star. 2015; at: http://www.thedailystar.net/op-ed/the-potential-solar-home-systems-
    bangladesh-82837, May 17, 2015 (6 September 2016).
UNFCCC. 2015. UNFCCC website on CDM; at: http://cdm.unfccc.int/ProgrammeOfActivities/
    poa_db/ZSI6WP0ODGRQ8UYKXB3MHTL957JVAE/view (6 September 2016).
World Bank. 2014; at: http://www.worldbank.org/en/news/press-release/2014/06/30/bangladesh-
    receives-usd-78-million-to-install-an-additional-480000-solar-home-systems (6 September 2016).

Chapter 13
Synthesis: A Future for Bangladesh Under a Changing Climate

Jeffrey Chow, Adrian Fenton, Saleemul Huq, Clare Stott, Julia Taub and Helena Wright

Abstract Due to its geophysical and socioeconomic characteristics, Bangladesh is particularly vulnerable to the harmful impacts of climate change. The contributors to this volume generally agree that climate projects should prioritize reducing exposure to loss and damage for the most at-risk populations, infrastructure, and assets. These include women, migrants, and the landless poor, whom would benefit from local-level projects that address their specific vulnerabilities. There are also a range of aspects, such as more robust monitoring, reporting, and evaluation, which require additional government commitment and policy integration. Confronting climate change will need to be continually responsive and adaptive to somewhat unpredictable future circumstances. With its experience and knowledge, Bangladesh has an important role to play as an example to the rest of the developing world.

Keywords Climate change · Bangladesh impacts · Future · Projections

Jeffrey Chow, International Centre for Climate Change and Development (ICCCAD), e-mail: Jchow.conservation@gmail.com.

Adrian Fenton, International Centre for Climate Change and Development (ICCCAD), Dhaka, Bangladesh.

Saleemul Huq, International Centre for Climate Change and Development (ICCCAD), Dhaka, Bangladesh.

Clare Stott, International Centre for Climate Change and Development (ICCCAD), Dhaka, Bangladesh.

Julia Taub, Global Network of Civil Society Organisations for Disaster Reduction (GNDR).

Helena Wright, International Centre for Climate Change and Development (ICCCAD), Dhaka, Bangladesh.

© Springer Nature Switzerland AG 2019
S. Huq et al. (eds.), *Confronting Climate Change in Bangladesh*,
The Anthropocene: Politik—Economics—Society—Science 28,
https://doi.org/10.1007/978-3-030-05237-9_13

13.1   Climate Change Impacts on Bangladesh

Bangladesh (20–26° N, 88–92° E) has geographical characteristics that render it one of the countries in the world most vulnerable to climate change-related damages. Most of the country is low-elevation and comprised of alluvial delta, with floodplains occupying 80% of its area (Agrawala et al. 2003). Located at the end of the Ganges-Brahmaputra-Meghna (GBM) river system, Bangladesh acts as a drainage outlet for cross-border runoff generated from China, Nepal, India, and Bhutan (Mirza 2005). The climate can be separated into four seasons: dry winter (December-February), pre-monsoon (March-May), monsoon (June-September), and post-monsoon (October-November) (Alamgir et al. 2015). Due to the high ratio of monsoon to dry season runoff, the country experiences an abundance of water during the monsoon season while also facing surface water scarcity in the dry season. Consequently, annual monsoon floods typically inundate one fifth to one third of the country (Faruque/Ali 2005)—and up to 70% in extreme years (Mirza 2003)—affecting both rural and densely populated urban areas. Occurring between June and September, these floods coincide with the main growing season when two-thirds of the total annual staple rice crop is produced. Although past severe floods fall within the range of historical variability, a warmer climate will likely increase precipitation and Himalayan glacier melt, resulting in greater frequency, magnitude, and extent of flooding and causing extensive damage to lives, property, and staple crops (Agrawala et al. 2003; Mirza 2011).

Despite being a flood-prone country, hydrological droughts are also common, particularly in northern and northwestern Bangladesh (Mirza 2005; Shahid/ Behrawan 2008; Alamgir et al. 2015). Dry season and pre-monsoon droughts typically affect the winter rice and wheat crops, resulting in greater irrigation demand (Faruque/Ali 2005). Drought can also occur when monsoon rains, which normally produce 80% of Bangladesh's annual precipitation, are significantly reduced or delayed, which severely impacts annual rice production (Agrawala et al. 2003; Faruque/Ali 2005). Though climate models generally predict an increase in monsoon and annual precipitation, increased drought during the dry winter season, exacerbated by higher evapotranspiration due to rising temperatures, remains a possibility.

The coastal zone in particular is extremely prone to flooding, as well as salinity intrusion, storm surges, and rapid geomorphological changes (Ali 1996, 1999; Brammer 2014). Bangladesh receives the brunt of cyclones, which originate in the Indian Ocean and funnel northward through the Bay of Bengal, where shallow waters contribute to tidal surges up to 15 m in height (Agrawala et al. 2003). While the intense precipitation generates inland and riverine flooding, storm surges cause most of the damage from tropical storms in Bangladesh, as surges that hit coastal areas propagate long distances inland along rivers (Ali 1996). Besides causing loss of life and damage to infrastructure, the storm surges ruin crops and aquaculture operations, which cannot tolerate saltwater intrusion. Coastal soil salinity is also exacerbated by high spring tides which inundate coastal areas and tidal flooding during the monsoon season (Karim et al. 1999; Haque 2006). Tropical storm surges —alongside heavy discharge currents through the GBM river system and wave action created by strong southwest monsoon winds—also contribute to coastal erosion (Ali 1999). Between 1973 and 2010, the Sundarbans coastline in southwestern Bangladesh underwent net erosion of approximately 6290 square km (Rahman et al. 2011), whereas to the east, the Tentulia, Meghna, and Feni River estuaries and nearby deltaic islands experienced about 2363 square km of accretion and 1337 square km of erosion during the same period (Chow 2017). Forming in warmer oceans, wind speed and precipitation from tropical cyclones are projected to intensify with climate change, increasing the damages caused by storm surge, floods, erosion, and salinity (IPCC 2013). Sea level rise—currently about 1.06 to 1.75 mm per year in the North Indian Ocean (Unnikrishnan/Shankar 2007), but up to 10.7 mm per year along the Bangladesh coast due to subsidence (Khan et al. 2000)—will aggravate these impacts, resulting in dramatically larger storm surge risk areas and flood depths over the course of the next century (Karim/Mimura 2008). As population growth results in more settlements and pressures on natural resources in the coastal zone, people and infrastructure will be increasingly exposed to climate hazards, especially among the poor (Brouwer et al. 2007; Brammer 2014).

Bangladesh's socioeconomic condition amplifies its vulnerability to climate impacts. Bangladesh is economically underdeveloped, with about a third of its population living in poverty, the highest rate in all South Asia (ADB 2017). Between 1991 and 2000, Bangladesh experienced 93 major natural disasters resulting in nearly 200,000 deaths and causing US$5.9 billion in damages with high losses in agriculture and infrastructure (MoEF GoB 2005). Although disaster-induced morbidity and mortality—often disproportionately affecting women (Neumayer/Plümber 2007)—have declined substantially over the last several decades thanks to improved early warning systems and shelters, economic losses may be increasing as the economy develops (Penning-Rowsell et al. 2013). Additionally, Bangladesh is an agrarian country primarily dependent on rice cultivation (BBS 2015), whose yields can be depressed by higher temperatures (Sarker et al. 2012; Chowdhury/Khan 2015), severe flooding (Paul/Rasid 1993; Banerjee 2010), and salinity (Dasgupta et al. 2014). Salinity in drinking water also results in public health problems, such as adverse maternal and fetal outcomes where pregnant women consume sodium well in excess of WHO-FAO recommended levels (Khan et al. 2011). Other possible long term health risks from climate change include greater incidences of heat-related mortality, water-borne infections, and vector-borne diseases (McMicheal et al. 2006). Heat and humidity extremes are expected to approach or exceed human survivability thresholds for some parts of Bangladesh by the late 21st century (Im et al. 2017). Thus, climate change may have severe consequences for public health, especially in light of the country's poor healthcare and sanitation infrastructure (Shadid 2010). Moreover, the temporary displacement and migration, which often follow climate disasters—as the landless poor seek shelter and employment away from their homes—are frequently accompanied by diminished livelihood opportunities and weakened social bonds (Penning-Rowsell et al. 2013).

13.2   Confronting Climate Change

Due to the range of adverse climate impacts, confronting climate change in Bangladesh has required actions across many diverse sectors. The Ministry of Environment and Forests (MoEF) within the Government of Bangladesh first outlined its adaptation strategy in its National Adaptation Programme of Action in 2005, which proposed mainstreaming adaptation into sectors ranging from education to infrastructure to forestry, while focusing particularly on disaster management, water, agriculture, and industry (MoEF GoB 2005). In 2009, the MoEF published the Bangladesh Climate Change Strategy and Action Plan (MoEF GoB 2009), which organized its climate change mitigation and adaptation goals into six pillars: food security; social protection and health; comprehensive disaster management; infrastructure; mitigation and low carbon development; research and knowledge management; and capacity building and institutional strengthening. The MoEF has also produced the Climate Change and Gender Action Plan (CCGAP), which integrates gender issues into the first four pillars and sets objectives to reduce the climate vulnerability of women (MoEF GoB 2013). Each of the six pillars emphasizes a strategic cross-sectoral approach. For example, Bangladesh intends to reduce greenhouse gas emissions not only through adopting low-carbon energy technology, but also by increasing carbon sequestration via community forestry. Alongside government efforts, local civil society organizations have also begun to contribute to adaptation planning.

As reviewed by Pervin et al. within this volume, adaptation programs are primarily undertaken by the public sector, with finance sourced from the Government of Bangladesh, multilateral development banks, and other sovereign donors. This funding is channeled through several windows, mainly the Pilot Program for Climate Resilience, the Bangladesh Climate Change Trust Fund, and the Bangladesh Climate Change Resilience Fund, which have allocated more than US $1.2 billion for over 400 projects across Bangladesh. The Government of Bangladesh has drafted a country investment plan to link the environment, forestry, and climate change sectors to the financing required for successful implementation (GoB 2017). This framework estimates US$4.9 billion in total funding is required for mitigation, adaptation, and resilience projects from 2016 to 2022, compared to the US$2.9 billion currently available.

Adaptations to protect food production from salinity and submergence, reviewed in this volume by Mondal et al., have required developing and promoting new, salt- and flood-resistant varietals of rice and non-rice crops, and bolstering infrastructure including embankments, drainage structures, and irrigation pumps and canals. Bangladesh has also explored innovative practices that reduce vulnerability by accommodating new inundation regimes, such as vertical horticulture, short-duration cropping, and alternating rotations of fish and crop cultivation. Some of these strategies are ecosystem-based, such that they incorporate biodiversity and ecosystem services into the adaptation program in ways that substitute or complement infrastructure and technology-based methods. In this volume, Saroar et al. identify key ecosystem-based adaptations for food production, which include rehabilitating natural canals and wetlands to enhance drainage and irrigation, as well as floating hydroponic agriculture and ditch-and-dyke schemes that integrate flood management with agriculture and aquaculture. Ecosystem-based adaptations may have multiple benefits, as well. For example, according to Chow et al. (this volume), the Government of Bangladesh has been implementing coastal mangrove conservation and afforestation not only to mitigate tropical storm surges and salinity intrusion, but also to use the binding properties of vegetation roots to reduce coastal erosion and enhance land accretion. Similarly, hillside afforestation has also been undertaken in order to mitigate slope erosion and landslides.

As reviewed by Mukherjee et al. in this volume, flood control, drainage, and irrigation have been the main goals of infrastructural adaptations implemented across Bangladesh. These have been utilized to reduce loss and damage from monsoon flooding in the north-west and north central region, tidal flooding in the eastern region, coastal flooding and storm surge in the southern coastal region, and water scarcity in the northwest and southwest regions. To reduce flood hazards for urban residents and infrastructure in the city of Dhaka, Bangladesh has invested heavily in river embankments, flood walls, pumping stations, sluice gates, as well as storm water drainage systems. As described by Rahman and Islam in this collection, the city established this protective infrastructure after a catastrophic flood in 1998, and continues to construct additional embankments and drainage works. Mitigating flooding has allowed former agricultural land in the urban fringe to be converted to residential zones. As Dhaka's population continues to grow, further non-structural measures have become necessary. Polyethylene bags have been banned in order to reduce congestion in the drainage system. Building codes and regulations now identify flood-prone areas and restrict large-scale developments in flood flow zones. However, illegal construction and settlements persist, and can impede drainage and endanger vulnerable residents, who tend to be poor.

Within the public health sector, the government, private sector, NGOs, and donor agencies have been working to improve access to healthcare services, as well as promote sanitary latrines. These organizations have also deployed various technologies and strategies for providing safe drinking water. Described by Rana in this volume, these include arsenic testing of groundwater, rainwater harvesting systems, pond sand filtration systems, and desalinization panels.

Bangladesh has also bolstered its communication systems for warning populations of hazardous climatic events such as floods and tropical storms. Within this collection, Afroz et al. discuss how telecasts of information from the Storm Warning Center have facilitated evacuations and reduced mortality. Rahman and Islam also describe the Flood Forecast and Warning Center, which provides live updates of river water levels and warnings when river water rises above safe levels, though this information is directed more towards decision-makers than the general public. The government has also increased general awareness of climate change through media campaigns and public education curricula and textbooks. Community-based organizations and NGOs play an important role in spreading climate knowledge and facilitating disaster preparedness in vulnerable areas.

In order to improve lives in Bangladesh's vulnerable rural areas, where two-thirds of the population lack access to grid electricity, a consortium comprised of the Government of Bangladesh, international donor agencies, NGOs, and private companies has been working to equip households with home solar electricity systems. More than 3 million systems have been installed, covering about 12 million people; however, this constitutes only a tenth of the rural population that could benefit from this technology. According to Muzammil and Ahmed (this volume), solar power adoption not only enhances household productivity and climate resilience, but also constitutes a low-carbon development strategy that reduces air pollution from the burning of kerosene.

As Reggers points out in her chapter, Bangladesh has made significant development gains in recent decades, doubling its Human Development Index score since 1980 and decreasing its poverty rate from 56.6% in 1991 to 31.5% in 2013. Despite the strides Bangladesh has made, there remain institutional and operational impediments that hinder the dissemination and adoption of adaptation strategies. Lack of adherence to flood zoning regulations and other governance failures result in a greater number of lives at risk. Insufficient investment capital and funding resources, a lack of knowledge, and coordination failures between government agencies, NGOs, and other stakeholders often render programs unsustainable. As discussed by Rana in this volume, health providers in particular often lack the funding and staff to ensure that comprehensive and specialized services are available to climate-vulnerable communities. There may be communication failures across institutions, since, as Afroz et al. mention in their chapter, the strategic framework for government ministries to interact on climate change is not clearly articulated in the Bangladesh Climate Change Strategic Action Plan. As Pervin et al. point out, civil society organizations often insufficiently communicate with each other and with public sector institutions, reducing their influence and constraining knowledge sharing. As a result, individual initiatives may be piecemeal and too narrowly focused on specific outcomes, when coordination and integration with other agencies and programs could potentially increase their effectiveness and probability of success. Some projects are also implemented in a top-down manner, disregarding consultations and input from local communities, and thus fail to account for their livelihood needs. One example, mentioned by Saroar et al. in this volume, is the Tidal River Management project, which has facilitated coastal land stabilization by retaining sediment, but has also reduced some fishing and cultivation opportunities for the local inhabitants.

While Bangladesh has been the recipient of myriad adaptation programs—making the country a laboratory for testing potential strategies—there has been a lack of rigorous follow-up evaluation. Consequently, program effectiveness often cannot be evaluated despite considerable uncertainty due to their frequently experimental nature. For example, Chow et al. point out within their chapter that the necessary tree plantation area or density to provide more than trivial mitigation of storm surges or landslides are largely unknown, and the paucity of reliable and systematic baseline data complicates ex post facto assessments. Well-intentioned projects in some cases may have immediate benefits but turn out to be maladaptive in the long run. As noted by Mukherjee et al., river embankments mitigate flood risks in the short term, but may over time trap sediments in the river bed, which reduces the replenishment of topsoil nutrients and increases future flood depth. Moreover, upstream flood embankments may increase geomorphological instability and flood risks downstream. Careful empirical analysis, research, monitoring, and evaluation are therefore necessary to determine the optimal deployment and management of adaptation strategies.

13.3 A Path Forward

To overcome these barriers, Bangladesh must develop long-term strategic plans, which build upon existing and potential synergies and avoid parallel and uncoordinated systems among all affected sectors. For instance, Saroar et al. recommend that ecosystem-based adaptation strategies must horizontally integrate forestry, fisheries, agriculture, and water resources in order to fully account for uncertainty and risk. Mukherjee et al. suggest that a single river basin is the appropriate unit of management and that one-size-fits-all, country-wide interventions that ignore local circumstances ought to be avoided. Thus, vertical coordination is essential as well, so that climate projects foster sustainability across various levels of implementation, while still taking into consideration the needs and practices of local communities, the private sector, and other relevant stakeholders. According to Pervin et al, finance mechanisms also could be more coordinated through the establishment of horizontal and vertical feedback systems and central information sharing hubs, in order to strengthen capacity, improve efficiency, and ensure future access to climate funds. Muzammil and Ahmed offer the example of solar home systems promotion in Bangladesh, which utilizes a holistic finance model that starts with public funding and policy, but also incorporates market development incentives, access to private capital, the establishment of delivery networks, after-sales services, and institutional support for partner organizations and small-scale enterprises.

There is general consensus among the contributors to this volume that climate projects should prioritize reducing the loss and damage exposure for the most at-risk populations, infrastructure, and assets. Such populations include women, migrants, and the landless poor, who would benefit from community-specific, local-level projects designed to address their particular vulnerabilities. More effective early warning systems and greater availability of disaster shelters can reduce loss of life, and local officials and media bear a responsibility to communicate with vulnerable people in order to improve hazard mitigation and response. Due to the heterogeneous characteristics of communities across Bangladesh, national strategies must anticipate and allow for flexible guidelines that are tailored to meet specific local needs. Local institutions such as village governments and NGOs ought to be afforded sufficient capacity, resources, and the authority to meaningfully contribute to adaptation projects within their jurisdictions.

Despite the considerable attention that has been given to adaptation in Bangladesh, there are aspects which require additional commitment by governments and integration into policy frameworks. Some impacts that span multiple jurisdictions, such as migration, displacement, and the disproportionate effect of climate change on women, merit greater attention and mainstreaming within national development plans. For example, Reggers reports that the CCGAP has yet to be implemented in any meaningful way at both national and local levels. Moreover, the dialogue concerning gender has focused almost exclusively on mitigating vulnerabilities, leaving inequality largely unaddressed. In urban areas, considerations of flood risks need to be embedded into city planning processes, since inadequate infrastructural maintenance, lack of enforcement against illegal settlements and commercial developments, and paucity of flood shelters place vulnerable populations at more risk, according to Rahman and Islam.

Stronger monitoring, evaluation, and reporting mechanisms are essential not only for transparent evaluation of climate projects, but also to review the performance of governance and finance. Afroz et al. also suggest that participatory monitoring could help increase resilience by spreading awareness of climate change, and that local communities can play an important role in identifying gaps. Moreover, robust evaluation would allow policy- and decision-makers to place more emphasis on effective strategies and to reduce support for ineffective ones, increasing the efficiency of climate fund expenditures. Increased confidence in the effectiveness of public financing could help encourage more private investment into climate projects, which would spread the responsibility and risks across a greater number of stakeholders. Mukherjee et al. suggest that one way of potentially increasing the cost-effectiveness of interventions would be to focus on projects with multiple, complementary benefits, particularly those that have both a emissions mitigation and adaptation effect. The elimination of ineffective strategies would also help avoid engendering a false sense of security among vulnerable populations. However, robust monitoring and evaluation would in many cases require long-term impact studies for which available data may be scarce. Scholars within academic institutions and NGOs have an important role to play in the rigorous and independent implementation of this research.

With decades of experience grappling with natural disasters that will increase in frequency and intensity due to climate change, Bangladesh has an important role to play as an exemplar to the rest of the developing world. Even if certain programs are unsuccessful in their initially defined adaptation or mitigation goals, analyzing the reasons for these failures may nevertheless offer lessons learned for future actions, improving climate strategies in the long run. If strategies prove to be cost-effective successes, then they help Bangladesh serve as a model for environmentally sound, socially responsible, and economically sustainable development in the face of climate change. Rigorous empirical evaluations are therefore vital, and are only possible if the Government of Bangladesh and other stakeholders are committed to transparency, large sample data collection, and most importantly knowledge sharing—the goal of this volume. Climate change-induced environmental hazards will only increase for the foreseeable future, against the backdrop of rapidly evolving socioeconomic conditions in Bangladesh. Confronting climate change in Bangladesh is therefore a task without determinable conclusions, and will need to be continually responsive and adaptive to somewhat unpredictable future circumstances.

Acknowledgments The authors would like to thank Yukyan Lam of the Natural Resources Defence Council and Sepul Barua of the Food and Agriculture Organization of the United Nations for reviewing and providing comments on this chapter.

References

Alamgir M, Shahid S, Hazarika Mk, Nashrrullah S, Harun Sb, Shamsudin S. 2015. Analysis of Meteorlogical Drought Pattern During Different Climatic And Cropping Seasons In Bangladesh. *Journal of the American Water Resources Association* 51: 794–806.

Ali A. 1996. Vulnerability of Bangladesh to climate change and sea level rise through tropical cyclones and storm surges. *Water, Air, and Soil Pollution* 92: 171–179.

Ali A. 1999. Climate change impacts and adaptation assessment in Bangladesh. *Climate Research* 12: 109–116.

Agrawala S, Ota T, Ahmed AU, Smith J, van Aalst M. 2003. *Development and Climate Change in Bangladesh: Focus on Coastal Flooding and the Sundarbans.* COM/ENV/EPOC/DCD/DAC (2003)3/FINAL. OECD.Asian Development Bank. 2017. Poverty in Bangladesh. https://www.adb.org/countries/bangladesh/poverty. Accessed 30 June 2017.

Banerjee L. 2010. Effects of flood on agricultural productivity in Bangladesh. Oxford *Development Studies* 38: 339–356.

Bangladesh Bureau of Statistics (BBS). 2015. *Yearbook of Agricultural Statistics-2013.* Statistics Division, Ministry of Planning: Dhaka, Bangladesh.

Brammer H. (2014). Bangladesh's dynamic coastal regions and sea-level rise. *Climate Risk Management* 1: 51–62.

Brouwer R, Akter S, Brander L, Haque E. 2007. Socioeconomic vulnerability and adaptation to environmental risk: a case study of climate change and flooding in Bangladesh. *Risk Analysis* 27: 313–326.

Chow J. 2017. Mangrove management for climate change adaptation and sustainable development of coastal zones. *Journal of Sustainable Forestry.* https://doi.org/10.1080/10549811.2017.1339615.

Chowdhury IUA and Khan MAE. 2015. The impact of climate change on rice yield in Bangladesh: a time series analysis. *Russian Journal of Agricultural and Socio-Economic Sciences* 4: 12–28.

Dasgupta S, Hossain MM, Huq M, Wheeler D. 2014. *Climate Change, Soil Salinity, and the Economics of High-Yield Rice Production in Coastal Bangladesh.* Policy Research Working Paper 7140. World Bank.

Faruque HSM and Ali ML. 2005. Climate change and water resources management in Bangladesh. In: M.M.Q. Mirza, Q.K. Ahmad (Eds.). *Climate Change and Water Resources in South Asia.* A.A. Balkema Publishers: London, UK. 231–254.

Government of Bangladesh (GoB). 2017. *Bangladesh Country Investment Plan on Environment, Forestry and Climate Change 2016–2021*. USAID, FAO, and Government of Bangladesh: Dhaka, Bangladesh.

Haque SA. 2006. Salinity problems and crop production in coastal regions of Bangladesh. *Pakistan Journal of Botany* 38: 1359–1365.

Im ES, Pal JS, Eltahir EAB. 2017. Deadly heat waves projected in the densely population agricultural regions of South Asia. *Science Advances* 3: e1603322.

Intergovernmental Panel on Climate Change (IPCC). 2013. Summary for Policymakers. In Stocker TF, Qin D, Plattner GK, Tignor MMB, Allen SK, Boschung J, Nauels A, Xia Y, Bex V, Midgley PM (eds), Climate Change 2013: *The Physical Science Basis. Contribution of Working Group I to the Fifth Assessment Report of the Intergovernmental Panel on Climate Change*. Cambridge University Press: Cambridge, United Kingdom and New York.

Karim Z, Hussain SG, Ahmed AU. 1999. Climate change vulnerability of crop agriculture. In: Huq S, Karim Z, Asaduzzaman M (eds). *Vulnerability and Adaptation to Climate Change for Bangladesh*. Springer Science + Business Media: Dordrecht, Netherlands. 39–54.

Karim MF, Mimura N. 2008. Impacts of climate change and sea-level rise on cyclonic storm surge floods in Bangladesh. *Global Environmental Change* 18: 490–500.

Khan AE, Ireson A, Kovats S, Mojumder SK, Khusru A, Rahman A, Vineis P. 2011. Drinking water salinity and maternal health in coastal Bangladesh: implications of climate change. *Environmental Health Perspectives* 119: 1328–1332.

Khan TMA, Singh OP, Rahman MS. 2000. Recent sea level and sea surface temperature trends along the Bangladesh coast in relation to the frequency of intense cyclones. *Marine Geodesy* 23 103–116.

McMichael AJ, Woodruff RE, Hales S. 2006. Climate change and human health: present and future risks. *Lancet* 367: 859–869.

Ministry of Environment and Forests Government of the People's Republic of Bangladesh (MoEF GoB). 2005. *National Adaptation Programme of Action*. Government of Bangladesh: Dhaka, Bangladesh.

Ministry of Environment and Forests Government of the People's Republic of Bangladesh (MoEF GoB). 2007. *Bangladesh Climate Change Strategy and Action Plan 2009*. Government of Bangladesh: Dhaka, Bangladesh.

Ministry of Environment and Forests Government of the People's Republic of Bangladesh (MoEF GoB). 2013. *Bangladesh Climate Change and Gender Action Plan*. Ministry of Environment of Forest, Government of the People's Republic of Bangladesh: Dhaka, Bangladesh.

Mirza MMQ. 2003. Three recent extreme floods in Bangladesh: A hydro-meteorological analysis. *Natural Hazards* 28: 35–64.

Mirza MMQ. 2005. The implications of climate change on river discharge in Bangladesh. In: Mirza MMQ, Ahmad QK (eds). *Climate Change and Water Resources in South Asia*. A.A. Balkema Publishers: London, UK. pp. 103–135.

Mirza MMQ. 2011. Climate change, flooding in South Asia and implications. *Regional Environmental Change* 11: S95–S107.

Neumayer E and Plümper T. 2007. The gendered nature of natural disasters: the impact of catastrophic events on the gender gap in life expectancy, 1981–2002. *Annals of the Association of American Geographers* 97: 551–566.

Paul BK and Rasid H. 1993. Flood damage to rice crop in Bangladesh. *Geographical Review* 83: 150–159.

Penning-Rowsell EC, Sultana P, Thompson PM. 2013. The 'last resort'? Population movement in response to climate-related hazards in Bangladesh. *Environmental Science & Policy* 27S: S44-S59.

Rahman AF, Dragoni D, El-Masri B. 2011. Response of the Sundarbans coastline to sea level rise and decreased sediment flow: a remote sensing assessment. *Remote Sensing of the Environment* 115: 3121–3128.

Sarker MAR, Alam K, Gow J. 2012. Exploring the relationship between climate change and rice yield in Bangladesh: an analysis of time series data. *Agricultural Systems* 112: 11–16.

Shahid S. 2010. Probably impacts of climate change on public health in Bangladesh. *Asia-Pacific Journal of Public Health* 22: 310–319.

Shahid S and Behrawan H. 2008. Drought risk assessment in the western part of Bangladesh. *Natural Hazards* 46: 391–413.

Unnikrishnan AS and Shankar D. 2007. Are sea-level-rise trends along the coasts of the north Indian Ocean consistent with global estimates? *Global and Planetary Change* 57: 301–307.

Correction to: Climate Governance and Finance in Bangladesh

Mousumi Pervin, Pulak Barua, Nuzhat Imam, Md. Mahfuzul Haque and Nahrin Jannat Hossain

Correction to:
Chapter 6 in: S. Huq et al. (eds.),
Confronting Climate Change in Bangladesh,
The Anthropocene: Politik—Economics—Society—Science 28, https://doi.org/10.1007/978-3-030-05237-9_6

In the original version of the book, the following belated corrections have been incorporated:

In chapter "Climate Governance and Finance in Bangladesh", the chapter author affiliations has been updated to

4. Md. Mahfuzul Haque: Transparency International Bangladesh, Dhaka, Bangladesh
5. Nahrin Jannat Hossain: Jagannath University, Dhaka, Bangladesh.

The updated version of this chapter can be found at
https://doi.org/10.1007/978-3-030-05237-9_6

© Springer Nature Switzerland AG 2019
S. Huq et al. (eds.), *Confronting Climate Change in Bangladesh*,
The Anthropocene: Politik—Economics—Society—Science 28,
https://doi.org/10.1007/978-3-030-05237-9_14

About the Institutions

International Centre for
Climate Change and
Development

The International Centre for Climate Change and Development (ICCCAD) is one of the leading research and capacity building organisations working on climate change and development in Bangladesh. ICCCAD's aim is to develop a world-class institution that is closely related to local experience, knowledge and research in one of the countries that is most affected by climate change. It is our mission to gain and distribute knowledge on climate change and, specifically, adaptation and thereby helping people to adapt to climate change with a focus on the global south. By focusing on such work in Bangladesh, ICCCAD allows international participants to gain direct knowledge of the issues in a real-world context.

Gobeshona is a knowledge sharing platform for climate change research on Bangladesh. It aims to bring together the national and international research community to encourage sharing, enhance quality and, in doing so, make climate change research on Bangladesh more effective.

© Springer Nature Switzerland AG 2019
S. Huq et al. (eds.), *Confronting Climate Change in Bangladesh*,
The Anthropocene: Politik—Economics—Society—Science 28,
https://doi.org/10.1007/978-3-030-05237-9

About the Editors

Dr. Saleemul Huq (Bangladesh) is the Director of the International Centre for Climate Change & Development (ICCCAD) since 2009. Dr. Huq is also a Senior Fellow at the International Institute for Environment & Development (IIED), where he is involved in building negotiating capacity and supporting the engagement of the Least Developed Countries (LDCs) in UNFCCC including negotiator training workshops for LDCs, policy briefings and support for the Adaptation Fund Board, as well as research into vulnerability and adaptation to climate change in the least developed countries. Dr. Huq has published numerous articles in scientific and popular journals, was a lead author of the chapter on Adaptation and Sustainable Development in the third assessment report of the Intergovernmental Panel on Climate Change (IPCC), and was one of the coordinating lead authors of 'Inter-relationships between adaptation and mitigation' in the IPCC's Fourth Assessment Report (2007). Prior to this, he was at Bangladesh Centre for Advanced Studies (BCAS) where he was in charge of management and strategy of the organisation. In 2000 he became an Academic Visitor at the Huxley School of Environment at Imperial College in London.

Email: saleemul.huq@iied.org
Website: https://www.iied.org/users/saleemul-huq

© Springer Nature Switzerland AG 2019
S. Huq et al. (eds.), *Confronting Climate Change in Bangladesh*,
The Anthropocene: Politik—Economics—Society—Science 28,
https://doi.org/10.1007/978-3-030-05237-9

Dr. Jeffrey Chow (Hong Kong S.A.R., China) is a graduate of the Yale University School of Forestry and Environmental Studies. While completing his PhD in Natural Resource and Environmental Economics, he was a Fulbright-Clinton Public Policy Fellow and Special Assistant to the Chief Conservator of Forests for the Bangladesh Forest Department. He has also served as a visiting scholar to BRAC and to the International Center for Climate Change and Development in Dhaka, as the Research Scientist and Technical Program Manager at Conservation International Hong Kong, and as a lecturer for the University of Hong Kong Department of Earth Sciences. His publications have appeared in Science, PLOS One, Land Economics, the Journal of Sustainable Forestry, and elsewhere. Dr. Chow is currently a consultant for Civic Exchange in Hong Kong. His research focuses on land economics and policy at the interface between forests and alternate uses such as agriculture and urban development, as well as on the role of natural capital in climate change adaptation.

Email: jchow.conservation@gmail.com

Dr. Adrian Fenton (Fiji/UK) is an adaptation specialist who is currently working as a National Advisor to the Government of Fiji, on the development of the National Adaptation Plan focusing on the integration of environmental and climate risk in development planning processes and the inclusion of the private sector. Previously, he worked as a consultant, completing projects for non-governmental organisations, research and policy institutes, and multilateral organisations such as the United Nations Development Programme. He holds a PhD from University of Leeds.

Email: adrianfenton@hotmail.co.uk
Website: https://uk.linkedin.com/in/adrianfenton

Clare Stott (UK) is a climate change consultant, working in the areas of adaptation, resilience and knowledge brokering. She holds an MSc in Anthropology, Environment and Development from University College London, where she specialised in climate change adaptation, exploring local level knowledge sharing in coastal Bangladesh. She subsequently worked with the International Centre for Climate Change and Development in Dhaka, Bangladesh, where she co-led the establishment of the Gobeshona platform. Following this, she was an independent consultant, delivering climate policy, governance and finance projects for the International Institute for Environment and Development. Clare is currently a consultant at Itad, working in the climate change theme to deliver monitoring, evaluation and learning assignments.

Email: crstott@hotmail.com

Julia Taub (UK) is the Project Officer at the Global Network of Civil Society Organisations for Disaster Reduction (GNDR), where she provides extensive support to a wide range of projects including Institutionalising Sustainable Community Based Disaster Risk Management and Views from the Frontline. With a Master's in International Development: Environment and Development from the University of Manchester, Julia spent a year working at the International Centre for Climate Change and Development (ICCCAD) in 2015–2016, where she was involved in organising CBA10 and helped set up a MEL programme. She has co-authored several publications and was the lead author of an article on loss and damage in the Paris Agreement.

Email: jetaub@mac.com

Dr. Helena Wright (UK) is a Senior Policy Advisor at environmental think tank E3G where she leads E3G's work on international financial institutions. Prior to this, Dr. Wright worked for the UK Government and the United Nations, as well as in several other roles focused on renewable energy and climate policy. She holds a PhD from Imperial College London, an MSc in Environmental Technology from Imperial, and degree in Social and Political Sciences from the University of Cambridge. Dr. Wright was a contributory author for the Intergovernmental Panel on Climate Change (IPCC) in its 5th Assessment report.

Email: drhelenawright@gmail.com
Website: https://uk.linkedin.com/in/helenawright

Printed in the United States
By Bookmasters